STATISTICAL ANALYSIS
IN MICROBIOLOGY:
STATNOTES

STATISTICAL ANALYSIS IN MICROBIOLOGY: STATNOTES

Richard A. Armstrong and Anthony C. Hilton

WILEY-BLACKWELL

A John Wiley & Sons, Inc. Publication

Published by John Wiley & Sons, Inc., Hoboken, New Jersey
Published simultaneously in Canada

For general information on our other products and services or for technical support, please contact our
Customer Care Department within the United States at (800) 762-2974, outside the United States at
(317) 572-3993 or fax (317) 572-4002.

Wiley also publishes its books in a variety of electronic formats. Some content that appears in print may not
be available in electronic formats. For more information about Wiley products, visit our web site at
www.wiley.com.

Library of Congress Cataloging-in-Publication Data is available

ISBN 978-0-470-55930-7

10 9 8 7 6 5 4 3 2 1

CONTENTS

Preface   xv

Acknowledgments   xvii

Note on Statistical Software   xix

1 ARE THE DATA NORMALLY DISTRIBUTED?   1

   1.1   Introduction   1

   1.2   Types of Data and Scores   2

   1.3   Scenario   3

   1.4   Data   3

   1.5   Analysis: Fitting the Normal Distribution   3

      1.5.1   How Is the Analysis Carried Out?   3

      1.5.2   Interpretation   3

   1.6   Conclusion   5

2 DESCRIBING THE NORMAL DISTRIBUTION   7

   2.1   Introduction   7

   2.2   Scenario   8

   2.3   Data   8

   2.4   Analysis: Describing the Normal Distribution   8

      2.4.1   Mean and Standard Deviation   8

      2.4.2   Coefficient of Variation   10

      2.4.3   Equation of the Normal Distribution   10

   2.5   Analysis: Is a Single Observation Typical of the Population?   11

      2.5.1   How Is the Analysis Carried Out?   11

      2.5.2   Interpretation   11

   2.6   Analysis: Describing the Variation of Sample Means   12

   2.7   Analysis: How to Fit Confidence Intervals to a Sample Mean   12

   2.8   Conclusion   13

3 TESTING THE DIFFERENCE BETWEEN TWO GROUPS   15

   3.1   Introduction   15

   3.2   Scenario   16

   3.3   Data   16

3.4   Analysis: The Unpaired t Test   16
       3.4.1   How Is the Analysis Carried Out?   16
       3.4.2   Interpretation   18
3.5   One-Tail and Two-Tail Tests   18
3.6   Analysis: The Paired t Test   18
3.7   Unpaired versus the Paired Design   19
3.8   Conclusion   19

4   WHAT IF THE DATA ARE NOT NORMALLY DISTRIBUTED?   21
4.1   Introduction   21
4.2   How to Recognize a Normal Distribution   21
4.3   Nonnormal Distributions   22
4.4   Data Transformation   23
4.5   Scenario   24
4.6   Data   24
4.7   Analysis: Mann–Whitney U test (for Unpaired Data)   24
       4.7.1   How Is the Analysis Carried Out?   24
       4.7.2   Interpretation   24
4.8   Analysis: Wilcoxon Signed-Rank Test (for Paired Data)   25
       4.8.1   How Is the Analysis Carried Out?   25
       4.8.2   Interpretation   26
4.9   Comparison of Parametric and Nonparametric Tests   26
4.10   Conclusion   26

5   CHI-SQUARE CONTINGENCY TABLES   29
5.1   Introduction   29
5.2   Scenario   30
5.3   Data   30
5.4   Analysis: 2×2 Contingency Table   31
       5.4.1   How Is the Analysis Carried Out?   31
       5.4.2   Interpretation   31
       5.4.3   Yates' Correction   31
5.5   Analysis: Fisher's 2×2 Exact Test   31
5.6   Analysis: Rows $\times$ Columns ($R \times C$) Contingency Tables   32
5.7   Conclusion   32

6   ONE-WAY ANALYSIS OF VARIANCE (ANOVA)   33
6.1   Introduction   33
6.2   Scenario   34
6.3   Data   34

6.4	Analysis		35
	6.4.1	Logic of ANOVA	35
	6.4.2	How Is the Analysis Carried Out?	35
	6.4.3	Interpretation	36
6.5	Assumptions of ANOVA		37
6.6	Conclusion		37

7 POST HOC TESTS   **39**

7.1	Introduction		39
7.2	Scenario		40
7.3	Data		40
7.4	Analysis: Planned Comparisons between the Means		40
	7.4.1	Orthogonal Contrasts	40
	7.4.2	Interpretation	41
7.5	Analysis: Post Hoc Tests		42
	7.5.1	Common Post Hoc Tests	42
	7.5.2	Which Test to Use?	43
	7.5.3	Interpretation	44
7.6	Conclusion		44

8 IS ONE SET OF DATA MORE VARIABLE THAN ANOTHER?   **45**

8.1	Introduction		45
8.2	Scenario		46
8.3	Data		46
8.4	Analysis of Two Groups: Variance Ratio Test		46
	8.4.1	How Is the Analysis Carried Out?	46
	8.4.2	Interpretation	47
8.5	Analysis of Three or More Groups: Bartlett's Test		47
	8.5.1	How Is the Analysis Carried Out?	47
	8.5.2	Interpretation	48
8.6	Analysis of Three or More Groups: Levene's Test		48
	8.6.1	How Is the Analysis Carried Out?	48
	8.6.2	Interpretation	48
8.7	Analysis of Three or More Groups: Brown–Forsythe Test		48
8.8	Conclusion		49

9 STATISTICAL POWER AND SAMPLE SIZE   **51**

9.1	Introduction		51
9.2	Calculate Sample Size for Comparing Two Independent Treatments		52
	9.2.1	Scenario	52
	9.2.2	How Is the Analysis Carried Out?	52

9.3   Implications of Sample Size Calculations   53
9.4   Calculation of the Power (P') of a Test   53
      9.4.1   How Is the Analysis Carried Out?   53
      9.4.2   Interpretation   54
9.5   Power and Sample Size in Other Designs   54
9.6   Power and Sample Size in ANOVA   54
9.7   More Complex Experimental Designs   55
9.8   Simple Rule of Thumb   56
9.9   Conclusion   56

10 ONE-WAY ANALYSIS OF VARIANCE (RANDOM EFFECTS MODEL): THE NESTED OR HIERARCHICAL DESIGN   57

10.1 Introduction   57
10.2 Scenario   58
10.3 Data   58
10.4 Analysis   58
      10.4.1 How Is the Analysis Carried Out?   58
      10.4.2 Random-Effects Model   58
      10.4.3 Interpretation   60
10.5 Distinguish Random- and Fixed-Effect Factors   61
10.6 Conclusion   61

11 TWO-WAY ANALYSIS OF VARIANCE   63

11.1 Introduction   63
11.2 Scenario   64
11.3 Data   64
11.4 Analysis   64
      11.4.1 How Is the Analysis Carried Out?   64
      11.4.2 Statistical Model of Two-Way Design   65
      11.4.3 Interpretation   65
11.5 Conclusion   66

12 TWO-FACTOR ANALYSIS OF VARIANCE   67

12.1 Introduction   67
12.2 Scenario   68
12.3 Data   68
12.4 Analysis   69
      12.4.1 How Is the Analysis Carried Out?   69
      12.4.2 Interpretation   70
12.5 Conclusion   70

13 SPLIT-PLOT ANALYSIS OF VARIANCE   **71**

13.1   Introduction   71
13.2   Scenario   72
13.3   Data   72
13.4   Analysis   73
    13.4.1   How Is the Analysis Carried Out?   73
    13.4.2   Interpretation   74
13.5   Conclusion   75

14 REPEATED-MEASURES ANALYSIS OF VARIANCE   **77**

14.1   Introduction   77
14.2   Scenario   78
14.3   Data   78
14.4   Analysis   78
    14.4.1   How Is the Analysis Carried Out?   78
    14.4.2   Interpretation   78
    14.4.3   Repeated-Measures Design and Post Hoc Tests   80
14.5   Conclusion   80

15 CORRELATION OF TWO VARIABLES   **81**

15.1   Introduction   81
15.2   Naming Variables   82
15.3   Scenario   82
15.4   Data   83
15.5   Analysis   83
    15.5.1   How Is the Analysis Carried Out?   83
    15.5.2   Interpretation   83
15.6   Limitations of r 85
15.7   Conclusion   86

16 LIMITS OF AGREEMENT   **87**

16.1   Introduction   87
16.2   Scenario   88
16.3   Data   88
16.4   Analysis   88
    16.4.1   Theory   88
    16.4.2   How Is the Analysis Carried Out?   89
    16.4.3   Interpretation   90
16.5   Conclusion   90

17 **NONPARAMETRIC CORRELATION COEFFICIENTS**   **91**

    17.1 Introduction   91

    17.2 Bivariate Normal Distribution   91

    17.3 Scenario   92

    17.4 Data   92

    17.5 Analysis: Spearman's Rank Correlation (ρ, r_s)   93

        17.5.1 How Is the Analysis Carried Out?   93

        17.5.2 Interpretation   94

    17.6 Analysis: Kendall's Rank Correlation (τ)   94

    17.7 Analysis: Gamma (γ)   94

    17.8 Conclusion   94

18 **FITTING A REGRESSION LINE TO DATA**   **95**

    18.1 Introduction   95

    18.2 Line of Best Fit   96

    18.3 Scenario   97

    18.4 Data   98

    18.5 Analysis: Fitting the Line   98

    18.6 Analysis: Goodness of Fit of the Line to the Points   98

        18.6.1 Coefficient of Determination (r^2)   98

        18.6.2 Analysis of Variance   99

        18.6.3 t Test of Slope of Regression Line   100

    18.7 Conclusion   100

19 **USING A REGRESSION LINE FOR PREDICTION AND
CALIBRATION**   **101**

    19.1 Introduction   101

    19.2 Types of Prediction Problem   101

    19.3 Scenario   102

    19.4 Data   102

    19.5 Analysis   102

        19.5.1 Fitting the Regression Line   102

        19.5.2 Confidence Intervals for a Regression Line   103

        19.5.3 Interpretation   104

    19.6 Conclusion   104

20 **COMPARISON OF REGRESSION LINES**   **105**

    20.1 Introduction   105

    20.2 Scenario   105

    20.3 Data   106

20.4   Analysis   106

    20.4.1   How Is the Analysis Carried Out?   106

    20.4.2   Interpretation   107

20.5   Conclusion   108

**21   NONLINEAR REGRESSION: FITTING AN EXPONENTIAL
CURVE   109**

21.1   Introduction   109

21.2   Common Types of Curve   110

21.3   Scenario   111

21.4   Data   111

21.5   Analysis   112

    21.5.1   How Is the Analysis Carried Out?   112

    21.5.2   Interpretation   112

21.6   Conclusion   112

**22   NONLINEAR REGRESSION: FITTING A GENERAL
POLYNOMIAL-TYPE CURVE   113**

22.1   Introduction   113

22.2   Scenario A: Does a Curve Fit Better Than a Straight Line?   114

22.3   Data   114

22.4   Analysis   114

    22.4.1   How Is the Analysis Carried Out?   114

    22.4.2   Interpretation   115

22.5   Scenario B: Fitting a General Polynomial-Type Curve   115

22.6   Data   116

22.7   Analysis   117

    22.7.1   How Is the Analysis Carried Out?   117

    22.7.2   Interpretation   117

22.8   Conclusion   118

**23   NONLINEAR REGRESSION: FITTING A LOGISTIC
GROWTH CURVE   119**

23.1   Introduction   119

23.2   Scenario   119

23.3   Data   120

23.4   Analysis: Nonlinear Estimation Methods   120

    23.4.1   How Is the Analysis Carried Out?   120

    23.4.2   Interpretation   121

23.6   Conclusion   122

24 NONPARAMETRIC ANALYSIS OF VARIANCE   **123**

24.1   Introduction   123
24.2   Scenario   123
24.3   Analysis: Kruskal–Wallis Test   124
    24.3.1   Data   124
    24.3.2   How Is the Analysis Carried Out?   124
    24.3.3   Interpretation   125
24.4   Analysis: Friedmann's Test   125
    24.4.1   Data   125
    24.4.2   How Is the Analysis Carried Out?   126
    24.4.3   Interpretation   126
24.5   Conclusion   126

25 MULTIPLE LINEAR REGRESSION   **127**

25.1   Introduction   127
25.2   Scenario   128
25.3   Data   128
25.4   Analysis   129
    25.4.1   Theory   129
    25.4.2   Goodness-of-Fit Test of the Points to the Regression
            Plane   131
    25.4.3   Multiple Correlation Coefficient (R)   131
    25.4.4   Regression Coefficients   131
    25.4.5   Interpretation   132
25.5   Conclusion   132

26 STEPWISE MULTIPLE REGRESSION   **135**

26.1   Introduction   135
26.2   Scenario   136
26.3   Data   136
22.4   Analysis by the Step-Up Method   136
    26.4.1   How Is the Analysis Carried Out?   136
    26.4.2   Interpretation   137
    26.4.3   Step-Down Method   137
26.5   Conclusion   138

27 CLASSIFICATION AND DENDROGRAMS   **139**

27.1   Introduction   139
27.2   Scenario   140
27.3   Data   140

27.4   Analysis   140
    27.4.1   Theory   140
    27.4.2   How Is the Analysis Carried Out?   142
    27.4.3   Interpretation   142
27.5   Conclusion   144

28   FACTOR ANALYSIS AND PRINCIPAL COMPONENTS ANALYSIS   145

28.1   Introduction   145
28.2   Scenario   146
28.3   Data   146
28.4   Analysis: Theory   147
28.5   Analysis: How Is the Analysis Carried Out?   148
    28.5.1   Correlation Matrix   148
    28.5.2   Statistical Tests on the Correlation Coefficient Matrix   148
    28.5.3   Extraction of Principal Components   149
    28.5.4   Stopping Rules   149
    28.5.5   Factor Loadings   149
    28.5.6   What Do the Extracted Factors Mean?   149
    28.5.7   Interpretation   150
28.6   Conclusion   152

References   153

Appendix 1   Which Test to Use: Table   157

Appendix 2   Which Test to Use: Key   159

**Appendix 3   Glossary of Statistical Terms and
Their Abbreviations   163**

**Appendix 4   Summary of Sample Size Procedures for
Different Statistical Tests   167**

Index of Statistical Tests and Procedures   169

PREFACE

This book is aimed primarily at microbiologists who are undertaking research and who require a basic knowledge of statistics to analyze their experimental data. Computer software employing a wide range of data analysis methods is widely available to experimental scientists. The availability of this software, however, makes it essential that investigators understand the basic principles of statistics. Statistical analysis of data can be complex with many different methods of approach, each of which applies in a particular experimental circumstance. Hence, it is possible to apply an incorrect statistical method to data and to draw the wrong conclusions from an experiment. The purpose of this book, which has its origin in a series of articles published in the Society for Applied Microbiology journal *The Microbiologist*, is an attempt to present the basic logic of statistics as clearly as possible and, therefore, to dispel some of the myths that often surround the subject. The 28 *statnotes* deal with various topics that are likely to be encountered, including the nature of variables, the comparison of means of two or more groups, nonparametric statistics, analysis of variance, correlating variables, and more complex methods such as multiple linear regression and principal components analysis. In each case, the relevant statistical method is illustrated with examples drawn from experiments in microbiological research. The text incorporates a glossary of the most commonly used statistical terms, and there are two appendices designed to aid the investigator in the selection of the most appropriate test.

Richard Armstrong and Anthony Hilton

ACKNOWLEDGMENTS

We thank the Society for Applied Microbiology (SFAM) for permission to publish material that originally appeared in *The Microbiologist*. We would also like to acknowledge Dr. Lucy Harper, the editor of *The Microbiologist*, for help in commissioning this book, supporting its production, and for continuing encouragement.

We thank Tarja Karpanen and Tony Worthington (both of Aston University) for the use of data to illustrate Statnotes 15, 18, and 20 and Dr. Steve Smith (Aston University) for the data to illustrate Statnote 13.

We thank Graham Smith (Aston University) for drawing the figures used in Statnotes 25 and 28.

This book benefits from the teaching, research data, critical discussion, and especially the criticism of many colleagues: Dr. T. Bradwell (British Geological Survey), Dr. N. J. Cairns (Washington University, St Louis), Dr. M. Cole (Aston University), Dr. R. Cubbidge (Aston University), Dr. C. Dawkins (University of Oxford), Dr. M. C. M. Dunne (Aston University), Dr. F. Eperjesi (Aston University), Professor B. Gilmartin (Aston University), Dr. I. Healy (King's College London), Dr. E. Hilton (Aston University), Dr. P. D. Moore (King's College London), Dr. S. N. Smith (Aston University), and Dr. K. M. Wade (University of Oxford).

We dedicate the book to our families.

NOTE ON STATISTICAL SOFTWARE

There are nearly 100 commercially available software packages for statistical analysis known to the authors at the time of this writing. At the present time, an authoritative review and comparison of all the available packages is not available and is beyond the scope of this book and its authors. However, the following points need to be considered. First, not all of the statistical tests most useful to microbiologists are likely to be available in a single package. Second, different packages may not use the same terminology when referring to various statistical tests. This problem is especially acute in describing the different forms of analysis of variance (ANOVA). A particular example of the confusion that can arise is discussed in Statnote 11 with reference to the terminology for the *two-way ANOVA in randomized blocks*. Third, it is better to become familiar with one or two packages than to have a superficial knowledge of many. Finally, although the authors do not offer any specific recommendations, the major software packages, for example, GRAPHPAD PRISM, MINITAB, SPSS, STATISTICA, MEDCALC, SYSTAT, UNISTAT, and STATVIEW will carry out most of the statistical procedures described in this book, including all the basic tests, most of the common variations of ANOVA (one way, two way, factorial, split plot, and repeated measures), contingency tables, nonparametric statistics including the comparison of more than two groups, correlation, and regression analyses including multiple regression. Despite the proliferation of statistical software and their accompanying manuals, sound advice from a statistician with knowledge of microbiology is likely to remain the best protection against the incorrect application of statistical procedures.

Statnote 1

ARE THE DATA NORMALLY DISTRIBUTED?

Why is knowledge of statistics necessary?

The role of statistics in an experimental investigation.

Types of data and scores.

Testing the degree of normality of the data: chi-square (χ^2) goodness-of-fit test or Kolmogorov–Smirnov (KS) test.

1.1 INTRODUCTION

Knowledge of statistical analysis is important for four main reasons. First, it is necessary to understand statistical data reported in increasing quantities in articles, reports, and research papers. Second, to appreciate the information provided by a statistical analysis of data, it is necessary to understand the logic that forms the basis of at least the most common tests. Third, it is necessary to be able to apply statistical tests correctly to a range of experimental problems. Fourth, advice will often be needed from a professional statistician with some experience of research in microbiology. Therefore, it will be necessary to communicate with a statistician, that is, to explain the problem clearly and to understand the advice given.

The scientific study of microbiology involves three aspects: (1) collecting the evidence, (2) processing the evidence, and (3) drawing a conclusion from the evidence. Statistical analysis is the most important stage of processing the evidence so that a sound

Statistical Analysis in Microbiology: Statnotes, Edited by Richard A. Armstrong and Anthony C. Hilton
Copyright © 2010 John Wiley & Sons, Inc.

conclusion can be drawn from the data. Two types of question are often posed by scientific studies. The first type of question is a test of a hypothesis, for example, does adding a specific supplement to a culture medium increase the yield of a microorganism? The answer to this question will be either yes or no, and an experiment is often designed to elicit this answer. By convention, hypotheses are usually stated in the negative, or as *null hypotheses* (often given the symbol H_0), that is, we prefer to believe that there is no effect of the supplement until the experiment proves otherwise. The second type of question involves the estimation of a quantity. It may be established that a particular supplement increases the yield of a bacterium, and an experiment may be designed to quantify this effect. Statistical analysis of data enables H_0 to be tested and the errors involved in estimating quantities to be determined.

1.2   TYPES OF DATA AND SCORES

There are many types of numerical data or scores that can be collected in a scientific investigation, and the choice of statistical analysis will often depend on the form of the data. A major distinction between variables is to divide them into parametric and nonparametric variables. When a variable is described as *parametric*, it is assumed that the data come from a symmetrically shaped distribution known as the normal distribution, whereas *nonparametric* variables have a distribution whose shape may be markedly different from normal and are referred to as *distribution free*, that is, no assumptions are usually made about the shape of the distribution.

In this book, three types of data are commonly collected:

1. Attribute data in which the data are frequencies of events, for example, the frequencies of males and females in a hospital with a particular infectious disease. In addition, frequency data can be expressed as a proportion, for example, the proportions of patients who are resistant to various antibiotics in a hospital or community-based environment.
2. Ranked data in which a particular variable is ranked or scored on a fixed scale, for example, the abundance of fungi in different soil environments might be expressed on a scale from 0 (none) to 5 (abundant).
3. Measurements of variables that fulfill the requirements of the normal distribution. Many continuous biological variables are normally distributed and include many measurements in microbiology. Not all measurements, however, can be assumed to be normally distributed, and it may be difficult to be certain in an individual case. The decision may not be critical, however, since small departures from normality do not usually affect the validity of many of the common statistical tests (Snedecor & Cochran, 1980). In addition, many parametric tests can be carried out if the sample size is large enough. It is worth noting that tests designed to be used on normally distributed data are usually the most sensitive and efficient of those available.

Statnote 1 is concerned with the basic question of whether the data depart significantly from a normal distribution and, hence, whether parametric or nonparametric tests would be the most appropriate form of statistical analysis.

1.3 SCENARIO

The domestic kitchen is increasingly recognized as an important reservoir of pathogenic microorganisms, with dishcloths and sponges providing an ideal environment for their growth, survival, and dissemination. Given the intrinsic structural and compositional differences between these two material types, a study was envisaged to investigate if one provided a more favorable environment for bacterial survival than the other, the hypothesis being that there would be a quantitative difference between the number of microorganisms recovered from dishcloths compared to sponges.

A total of 54 "in-use" dishcloths and 46 sponges were collected from domestic kitchens, and the aerobic colony count of each was determined in the laboratory. Microbiological data from environmental sources usually have very large counts and associated standard deviations (SD) (see Statnote 2) and may not be normally distributed. Hence, the first stage of the analysis, illustrated by the sponge data in this statnote, is to determine the degree to which, if at all, the data depart from normality. Having established the distribution of the data, parametric or nonparametric statistical tests may be applied as appropriate to compare the difference between the cloth and sponge data.

1.4 DATA

The data comprise the aerobic colony counts of bacteria on 46 sponges and therefore represent several measurements of a single variable.

1.5 ANALYSIS: FITTING THE NORMAL DISTRIBUTION

1.5.1 How Is the Analysis Carried Out?

To fit the normal distribution, the variable (aerobic colony count on sponges) is first divided into frequency classes describing the range of the variable in the population. In the present case, 10 classes were used (Table 1.1). The limits that define these classes are then converted so that they are members of the *standard normal distribution*, that is, a distribution that has a mean of zero and a SD of unity. To carry out this calculation, the mean and SD of the observations are first calculated. The sample mean is then subtracted from each class limit and divided by the SD, which converts the original measurements to those of the standard normal distribution. Tables of the standard normal distribution are then used to determine the expected number of observations that should fall into each class if the data are normally distributed. The observed and expected values (Fig. 1.1) are then compared using either a chi-square (χ^2) "goodness-of-fit" or a Kolmogorov–Smirnov (KS) test. This statistical analysis is available in most of the popular statistical analysis software packages.

1.5.2 Interpretation

The χ^2 test ($\chi^2 = 38.99$) for the sponge data is significant at the 1% level of probability ($P < 0.01$), suggesting that the data deviate significantly from a normal distribution. The KS test (KS = 0.0895), however, is not significant ($P > 0.05$), a not uncommon result since this test is less sensitive than χ^2 and only indicates gross deviations from the normal

TABLE 1.1   Observed and Expected Frequencies for Sponge Data[a,b]

Category (Upper Limits)	O	E	Deviation $O - E$
<=5000000	6	5.78	0.22
60000000	11	7.91	3.09
115000000	8	10.864	−2.86
170000000	14	10.33	3.67
225000000	4	6.795	−2.795
280000000	1	3.09	−2.09
335000000	0	0.98	−0.97
390000000	1	0.21	0.79
445000000	1	0.03	0.967
<Infinity	0	0.003	−0.003

1. Divide the variable into frequency classes describing the range of the variable in the population. In the present case, 10 classes were used.
2. The limits that define these classes are then converted so that they are members of the standard normal distribution: *New class limit = (original class limit − mean)/standard deviation (SD).*
3. Tables of the standard normal distribution are used to determine the expected number of observations (E) that should fall into each class if the data are normally distributed.
4. Compare the observed (O) and expected (E) values using either a chi-square (χ^2) goodness-of-fit or a Kolmogorov–Smirnov (KS) test.

[a] Tests of normality: χ^2 (all categories) = 38.99 ($P < 0.01$); χ^2 (adjusted for expected values <5) = 4.80 ($P > 0.05$); KS test = 0.0894 ($P > 0.05$).
[b] O = Observed frequency, E = expected frequency as predicted by the normal distribution.

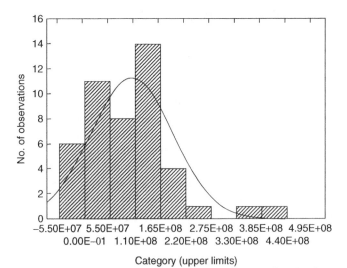

Figure 1.1. Histogram illustrating the observed distribution of values for the sponge data and the predicted normal distribution (continuous line). The chi-square (χ^2) goodness-of-fit and Kolmogorov–Smirnov (KS) tests test the difference between the observed and expected frequencies. Because of the large absolute counts, the limits of each class are in scientific units, e.g., E7 represents 1×10^7.

distribution (Pollard, 1977). The χ^2 test also has limitations as it is affected by how many categories are selected to define the variable and how these categories are divided up. In addition, if a number of the categories have expected numbers of observations below 5, adjacent categories should be combined until their expected values are greater than 5. If this procedure is carried out using the present data, the value of χ^2 is not significant ($\chi^2 = 4.80$, $P > 0.05$). In such situations, the general shape of the observed distribution is probably the best method of judging normality. Although this distribution (Fig. 1.1) exhibits a degree of skew (see Statnote 4), the deviations from normal (supported by the KS test) are not significant enough to warrant using a nonparametric test. However, a similar analysis carried out on the cloth data resulted in much greater deviations from a normal distribution as shown on both tests ($\chi^2 = 3007.78$, $P < 0.001$; $KS = 0.28$, $P < 0.01$). Hence, taking all these considerations into account, it may be prudent not to use parametric statistics (in this case an unpaired t test; see Statnote 3) to compare the cloth and sponge data. There are two ways to proceed: (1) transform the data to a new scale that is normally distributed, thus allowing the application of the parametric "unpaired" t test (see Statnote 3), or (2) to employ the nonparametric equivalent of the unpaired t test, namely the Mann–Whitney test (see Statnote 4).

1.6 CONCLUSION

Testing whether an observed distribution of observations deviates from a normal distribution is a common statistical procedure and available in many statistical packages. Most statistical software will offer two methods of judging whether there are significant deviations of the observed from the expected distributions, namely the χ^2 and the KS tests. These tests have different sensitivities and limitations and may give conflicting results. Hence, the results of these tests together with observations of the shape of the observed distribution should be used to judge normality.

Statnote 2

DESCRIBING THE NORMAL DISTRIBUTION

The normal distribution (mean and standard deviation).
Is an individual observation typical of a population?
Variation of a sample mean (standard error of the mean).
Confidence interval of a sample mean.

2.1 INTRODUCTION

In Statnote 1, two procedures, namely the chi-square (χ^2) and the Kolmogorov–Smirnov (KS) tests, were described to determine whether a sample of data came from a normal distribution. If the sample of observations do come from a normal distribution, then a number of statistics can be calculated that describe the central tendency (*mean*) and degree of spread (*standard deviation*) (SD) of the sample. In addition, a sample of measurements can be used to make inferences about the mean and spread of the population from which the sample has been drawn. This statnote describes the application of the normal distribution to some common statistical problems in microbiology, including how to determine whether an individual observation is a typical member of a population and how to obtain the *confidence interval* of a sample mean.

Statistical Analysis in Microbiology: Statnotes, Edited by Richard A. Armstrong and Anthony C. Hilton
Copyright © 2010 John Wiley & Sons, Inc.

TABLE 2.1   Dry Weight of Bacterial Biomass under Unsupplemented (UNS) and Supplemented (S) Growth Conditions in a Sample of 25 Fermentation Vessels

UNS	S	UNS	S	UNS	S
461	562	506	607	518	617
472	573	502	600	527	622
473	574	501	603	524	626
481	581	505	605	529	628
482	582	508	607	537	631
482	586	500	609	535	637
494	591	513	611	542	645
493	592	512	611		
495	592	511	615		

2.2   SCENARIO

An experiment was carried out to investigate the efficacy of a novel media supplement in promoting the development of cell biomass. Two 10-liter fermentation vessels were sterilized and filled with identical growth media with the exception that the media in one of the vessels was supplemented with 10 ml of the novel compound under investigation. Both vessels were allowed to equilibrate and were subject to identical environmental/incubation conditions. The vessels were then inoculated with a culture of bacterium at an equal culture density and the fermentation allowed to proceed until all the available nutrients had been exhausted and growth had ceased. The entire volume of culture medium in each fermentation vessel was then removed and filtered to recover the bacterial biomass, which was subsequently dried and the dry weight of cells measured.

2.3   DATA

The experiment was repeated 25 times and the dry weight of biomass produced in each experiment was recorded in Table 2.1. This statnote will only be concerned with analysis of the data from the supplemented culture, while in Statnote 3 the same data will be used to determine the significance of the difference between media with and without supplement. As in Statnote 1, the data comprise a sample of measurements of a single variable.

2.4   ANALYSIS: DESCRIBING THE NORMAL DISTRIBUTION

2.4.1   Mean and Standard Deviation

If the sample of measurements of bacterial biomass ($N = 25$) on supplemented media (X) is plotted as a *frequency distribution* (Fig. 2.1), the measurements appear to be more or less symmetrically distributed around a central tendency or average value. If the measurements were to be increased to a very large number and the class intervals of the distribution reduced to zero, the data would approximate closely to a continuous, bell-shaped curve called the normal distribution (also known as a Gaussian distribution). Many measurements in the biosciences follow this type of distribution or do not deviate significantly

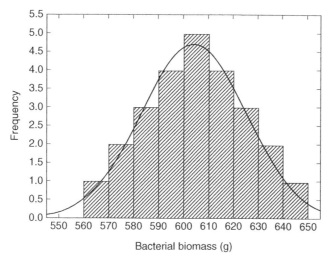

Figure 2.1. Frequency distribution of bacterial biomass on supplemented media. The curve shows the normal distribution fitted to these data. (Goodness-of-fit test: KS = 0.079, $P > 0.05$.)

from it. In the present case, the sample data did not deviate significantly from normal as indicated by a KS test, and we will proceed on the assumption that the data are normally distributed (see Statnote 1).

The normal distribution can be described by two statistics:

1. The average or "arithmetic mean" (μ) of the population:

$$\mu = \frac{\sum x}{n} \tag{2.1}$$

where x stands for each item in the sample taken successively; all observations being summed (Σ) and divided by n, the number of observations. Note that the mean of a sample of measurements taken from this population will be shown as x^*.

2. The SD of the population, that is, the distance from the mean to the point of maximum slope on the curve. The SD describes how closely the data cluster around the mean. Note that the SD of a population is given by the symbol σ while that of a sample from the population will be shown as s. The SD of a population of observations is given by

$$\sigma = \frac{\sqrt{\sum (x - \mu)^2}}{n} \tag{2.2}$$

To calculate the SD requires μ, the mean of the population, to be known. In most circumstances, however, it is the calculation of the SD of a small sample of measurements that is required. In the present example, the exact value of μ is not known but the sample mean x^* can be substituted. Hence, to calculate the SD of a sample of measurements, Eq. (2.2) can be used but incorporating three changes:

1. The SD of the population σ is replaced by the symbol s, the SD of the sample.
2. Mean μ is replaced by x^*, the mean of the sample.
3. Quantity n is replaced by $n - 1$, a quantity called the *degrees of freedom* (DF).

To understand the concept of DF, it is useful to note that the calculation of the SD involves the subtraction of individual observations from their mean, which are then squared and summed. However, if there are n observations in the sample, once $n - 1$ observations have been subtracted from the mean, the last deviation can be calculated from the existing information because the sum of all of the deviations from the mean must be zero. In other words, n observations only provide $n - 1$ independent estimates of the deviations from the mean. As a general rule, the DF of a statistical quantity is the number of "independent" observations making up that quantity.

Hence, the formula for the SD of a sample is

$$s = \frac{\sqrt{\sum (x - x^*)^2}}{n - 1} \tag{2.3}$$

If several estimates of the same quantity are made in a study, it is common practice to report the mean and the SD of the sample. In the present example, we would describe the sample of biomass measurements on supplemented media as having a mean of 604.28 and an SD of 21.16.

2.4.2 Coefficient of Variation

Another useful way of expressing the variability of a sample is as the *coefficient of variation* (CV), that is, the SD expressed as a percentage of the mean:

$$CV = \frac{s \times 100}{x^*} \tag{2.4}$$

The CV provides a standardized method of expressing the variability of a measurement in an experiment. Each variable in microbiology often has a characteristic CV that is relative constant across similar experiments, and therefore, it may be possible to obtain an estimate of this variability in advance by examining the results of previous experiments. The CV is therefore useful in the design of experiments. In the present case, the CV for the supplemented data is 3.5%.

2.4.3 Equation of the Normal Distribution

The mathematical equation that describes the normal distribution is given as follows (Snedecor & Cochran 1980):

$$y = \frac{1}{\sigma\sqrt{2\pi}} e^{-(x-\mu)^2/2\sigma^2} \tag{2.5}$$

This equation enables the height of the normal curve (y) to be calculated for each individual value of x, providing that μ and σ of the distribution are known. This equation also enables the proportion of observations that fall a given distance from the mean to be calculated.

For example, in any normal distribution, approximately 68% of the observations will fall one SD above and below the mean. Hence, the probability is 68% or $P = 0.68$ that a single measurement from a normal distribution will fall between these limits. Similarly, the probability is $P = 0.95$ that a single measurement will fall approximately two SD above and below the mean. Each type of variable in microbiology will have a characteristic normal distribution of values with a typical mean and SD. Statistical tables of the normal distribution, called z tables, however, have been calculated for a single distribution termed the *standard normal distribution*. If tables of the standard normal distribution are used in statistical tests, then measurements have to be converted so that they become members of the standard normal distribution.

2.5 ANALYSIS: IS A SINGLE OBSERVATION TYPICAL OF THE POPULATION?

2.5.1 How Is the Analysis Carried Out?

The standard normal distribution has a mean of zero ($\mu = 0$) and a SD of one unit ($\sigma = 1$) and provides the basis of many useful statistical tests. For example, it may be important to determine whether a single "new" observation x is typical or atypical of a population of measurements. To make this test, the original observation x has to be converted so that it becomes a member of the standard normal distribution z:

$$z = \frac{\pm(x - \mu)}{\sigma} \tag{2.6}$$

Tables of the standard normal distribution can then be used to determine where z is located relative to the mean of the distribution, that is, does it fall near the mean of the distribution (a typical value) or out in one of the tails of the distribution (an atypical value). An important question is how atypical does z have to be before it could be considered atypical of the population? By convention, x is considered to be a typical member of the population unless it is located in the tails of the distribution, which include the 5% most extreme values. The value of z that separates the typical values (95% of the distribution) from the atypical values (5% of the distribution, 2.5% in each tail of the distribution) is actually 1.96. Hence, if our calculated value of z is equal to or greater than 1.96, we would consider the measurement to be atypical.

2.5.2 Interpretation

As an example, assume that we made an additional estimate x of bacterial biomass under supplemented conditions and obtain a value of 550. Is this value typical of the "population" of values defined in Figure 2.1? To carry out the test, subtract the mean from x and divide by the SD to convert x to z. A value of $z = -2.56$ was obtained, which is greater than 1.96, the value of z that cuts off the 5% most extreme observations in the population. Hence, a bacterial biomass of 550 is not typical of the values obtained previously under the defined conditions, and there would be some doubt as to whether the conditions of the original experiment had been exactly reproduced in making the additional estimate of x.

2.6   ANALYSIS: DESCRIBING THE VARIATION OF A SAMPLE MEAN

If the study on supplemented media was repeated with several samples of 25 flasks, the same mean value would not necessarily be obtained each time, that is, the means of samples also vary. Hence, we may wish to assess how good an estimate our individual sample mean was of the actual population mean. To answer this question requires knowledge of how means from a normal distribution of individual measurements themselves vary. To understand this concept, it is necessary to quote an important statistical result termed the *central limit theorem*. This states that means from a normal distribution of individual values are themselves normally distributed with a mean of μ and an SD of $s/\sqrt{n}$, where n is the number of observations in the sample. In addition, the means of many nonnormal distributions will be normally distributed as long as the samples are large enough. It is important to distinguish the quantity $s/\sqrt{n}$, the SD of the population of sample means or *standard error of the mean* (SE) from σ or s the SD of a population or sample of individual measurements.

2.7   ANALYSIS: HOW TO FIT CONFIDENCE INTERVALS TO A SAMPLE MEAN

The SE of the mean is often used to plot on a line graph a *confidence interval* (CI) or error bar, which indicates the degree of confidence in the sample mean as an estimate of the population mean. The CIs are calculated as follows:

1. If a single observation x comes from a normal distribution, then the probability is 95% ($P = 0.95$) that x will be located between the limits $\mu \pm 1.96\sigma$.
2. Similarly, if a sample mean $x*$ comes from a normal population of sample means, then $P = 0.95$ such that $x*$ lies between the limits $\mu \pm 1.96\sigma/\sqrt{n}$.
3. Hence, $P = 0.95$ such that μ lies between the limits:

$$x* \pm \frac{1.96\sigma}{\sqrt{n}} \tag{2.7}$$

There are two problems with this calculation. First, in the majority of studies, the sample mean $x*$ is based on a small sample of measurements. Hence, the value of σ is not known, only the SD of the sample s, and we, therefore, substitute s for σ. Second, if n is small, the exact shape of the underlying distribution is uncertain, and therefore we cannot be sure whether the value of $Z = 1.96$ is accurate enough to judge whether a sample mean is atypical of the population. Instead, a different value is used that more accurately describes the behavior of small samples, namely a value from the t distribution. The t distribution will be discussed in more detail in the Statnote 3.

4. Hence, the 95% CI of a sample mean are given as

$$\text{CI} = x* \pm t\,(P = 0.05, \text{DF} = n-1)\left(s/\sqrt{n}\right) \tag{2.8}$$

For the supplemented biomass data, the 95% CI were estimated to be 604.28 ± 8.72.

Therefore, it can be stated with 95% confidence that the population mean falls between the calculated limits. The 95% CIs are often plotted as error bars on a line graph. It is important to understand what the error bars represent since investigators may plot the SD of a sample, the SE of the sample mean, or the 95% CI, and each conveys different information. In addition, error bars should not be used to make judgments as to whether there are significant differences between two or more of the means represented on the graph. The CI of two sample means are calculated using the SE appropriate to those sample means alone. To test whether the two means are different requires another form of SE, that is, the *standard error of the difference between two means*, and this will be discussed in Statnote 3.

2.8 CONCLUSION

If a sample of measurements comes from a population that is normally distributed, several statistics can be used to describe the sample, such as the mean, SD, and CV. In addition, how atypical an individual measurement has to be before it would be considered not to be a member of a specific population can be tested. Furthermore, the sample can be used to make inferences about the population from which the sample was drawn, namely estimating the population mean and by fitting 95% CI.

Statnote 3

TESTING THE DIFFERENCE BETWEEN TWO GROUPS

Distribution of the differences between pairs of sample means.
Comparing the difference between two means: the t test.
One-tail and two-tail tests.
Paired and unpaired t tests.

3.1 INTRODUCTION

Statnote 1 described a statistical test to determine whether a sample of measurements came from a normal distribution. If a variable is normally distributed, it is referred to as a *parametric* variable. A sample of observations from a parametric variable can be described by their sample mean x^* (the central tendency of the observations) and the standard deviation (SD) s (the degree of variability or spread of the observations) (see Statnote 2). Two statistical procedures based on the normal distribution were also described. First, a test of whether an individual measurement was typical or atypical of a larger population. Second, it was shown that the mean of a small sample of measurements is also a member of a normal distribution, namely the population of sample means. The degree of spread of this distribution can be described by the SE of the mean, and this information was used to calculate a 95% confidence interval for a sample mean. In this Statnote, these statistical ideas are extended to the problem of testing whether there is a statistically significant difference between two samples of measurements.

Statistical Analysis in Microbiology: Statnotes, Edited by Richard A. Armstrong and Anthony C. Hilton

3.2 SCENARIO

We return to the scenario described in Statnote 2, which investigated the efficacy of a novel media supplement in promoting the development of cell biomass. To recapitulate, two 10-liter fermentation vessels were filled with identical growth media, with the exception that the media in one of the vessels was supplemented with 10 ml of a novel compound. Both vessels were allowed to equilibrate and were subject to identical conditions. The vessels were then inoculated with a culture of bacteria and the fermentation allowed to proceed until bacterial growth had ceased. The entire volume of culture media in each fermentation vessel was then removed and filtered to recover the bacterial biomass. The experiment was repeated 25 times.

3.3 DATA

The data comprise two groups or treatments, namely with and without the novel supplement added, and there were 25 repeats or replicate measurements of each group. The group means together with their CIs are shown in Figure 3.1. The raw biomass data are presented in Table 2.1.

3.4 ANALYSIS: THE UNPAIRED *t* TEST

3.4.1 How Is the Analysis Carried Out?

To determine whether an individual measurement is typical of a population requires knowledge of how individual measurements vary, that is, the SD of the population (Statnote 2). Similarly, to determine the degree of error associated with a sample mean requires

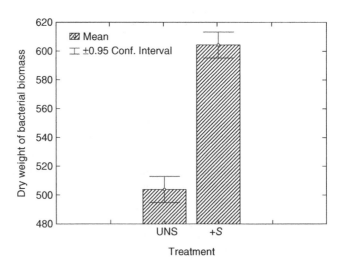

Figure 3.1. Mean bacterial biomass (with 95% CIs) using unsupplemented (UNS) and supplemented (+S) media (*t* test for comparing the means of both groups, $t = 16.59$, $P < 0.001$). Note that the confidence intervals should *not* be used as a basis for judging whether there is a significant difference between the means.

knowledge of how means vary, that is, the SE of the mean (Statnote 2). Extrapolating further: To determine whether there is a significant difference between the means of two samples, knowledge is required of how the differences between two sample means would vary. Hence, for each of the two groups, the mean is calculated and the mean of the unsupplemented (UNS*) group subtracted from the mean of the supplemented (S*) group to obtain the difference between the means. This difference represents an estimate of the "treatment effect" of the experiment, that is, the degree to which the media supplement may have increased bacterial biomass. If this experiment was exactly repeated several times under the same conditions, several estimates of UNS* − S* would be obtained, and a frequency distribution of the differences between the means could be constructed. If the distributions of the means from the supplemented and unsupplemented groups are themselves normally distributed, then the distribution of the differences between pairs of means taken from these two populations is also likely to be normally distributed. As a result, the standard normal distribution can be used to test whether there is a true difference between these means.

The means of the two treatments differed by 100.16 units (Fig. 3.1), and the two samples exhibited relatively little overlap (Fig. 3.2), suggesting a real effect of the supplement. There is, however, variation in microbial biomass between the replicate flasks within each group. Hence, is the difference between the means attributable to the effect of the supplement or could it be accounted for by random variation between the flasks? To decide between these two alternatives, the treatment effect ($U* - S*$) is compared with the degree of variation between the individual measurements, which has been pooled from both groups, by carrying out a *t* test. The statistic *t* is the ratio of the difference between the two means to the SE of the difference between the means:

$$t = \frac{U* - S*}{\sqrt{s_1^2/n_1 + s_2^2/n_2}} \tag{3.1}$$

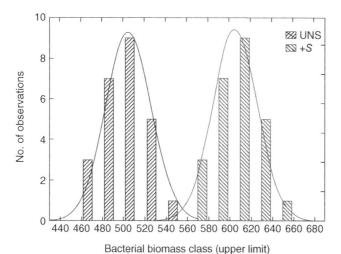

Figure 3.2. Frequency distribution of the individual biomass measurements using unsupplemented (UNS) and supplemented (+S) media. Curves represent the normal distribution fitted to each sample.

where s_1 and s_2 are the variances of each group and n_1 and n_2 the number of observations within each group ($n_1 = n_2 = 25$). Note that this calculation is similar to that carried out in Statnote 2 in which an individual value x was converted so that it became a member of the standard normal distribution. In the present example, the t distribution is used instead of the standard normal distribution because t describes the variation of means calculated from relatively small numbers of observations more accurately. Hence, when t is calculated, the difference between the means ($U^* - S^*$) becomes a member of the t distribution. The probability of obtaining a value of t of this magnitude by chance from two groups of samples when each comprises 25 observations is obtained either from statistical software or by consulting statistical tables (Fisher & Yates, 1963).

3.4.2   Interpretation

In the present example, $t = 16.59$, and this value was taken to a statistical table of the t distribution, entering the table for $n_1 + n_2 - 2$ DF (DF = 48), that is, the total number of observations minus 1 from each column. When t is equal to or greater than 2.01 (the value at $P = 0.05$ at 48 DF), the value is in a region of the distribution that includes the 5% most extreme values. Hence, $t = 16.59$ is an unlikely value to occur by chance; in fact the probability of obtaining this value by chance is less than 1 in a 1000 ($P < 0.001$), and, therefore, there is a real difference between the two means.

3.5   ONE-TAIL AND TWO-TAIL TESTS

Given the present scenario, it is possible to test two different H_0. First, it can be hypothesized that the addition of the supplement (S) would have no effect on bacterial biomass. This hypothesis does not specify whether a significant increase or a decrease in biomass would be necessary to refute the H_0. In this case, a *two-tailed test* would be appropriate, that is, both tails of the t distribution are used to test the H_0. Second, it could be hypothesized that the supplement would only increase biomass since it may be known in advance that it could not be significantly decreased by the treatment. If the hypothesis specifies whether a positive or a negative effect is necessary to refute the hypothesis, a *one-tail test* would be appropriate. Some statistical tables indicate both the one-tail and two-tail probabilities corresponding to particular columns. Most statistical tables, however, with some notable exceptions, only indicate the two-tail probabilities. To find the one-tail probability in a two-tail table, halve the probability; that is, the 5% one-tail probabilities are found in the 10% two-tail column.

3.6   ANALYSIS: THE PAIRED t TEST

An experiment involving two treatments or groups can be carried out by two different methods, namely the *unpaired* (independent) and the *paired* (dependent) methods. The experiment described in the scenario was carried out using an unpaired design; that is, the media supplement was allocated at random and without restriction to half of the 50 original flasks. In a paired design, however, the 50 flasks would be first divided into 25 pairs with the intention of processing a single pair (one supplemented, the other unsupplemented) on each of 25 separate days. Second, the supplement would then be allocated to one flask of each pair independently and at random. Hence, there is now a restriction in the method of allocation of the treatments to the flasks and a different analysis is required. In a paired design, the t test is defined as follows:

$$t = \frac{d*}{s_d/\sqrt{n}}$$
(3.2)

In this case, $d*$ is the mean of the differences between each of the 25 pairs of observations and s_d is the SD of these differences. The same t table is used for determining the significance of t. In a paired t test, however, a different rule is used for entering the t table; namely t has $n - 1$ DF, where n is the number of pairs of subjects. One-tail or two-tail tests may be made as appropriate.

3.7 UNPAIRED VERSUS THE PAIRED DESIGN

Is a paired or an unpaired design the best method of carrying out the proposed experiment? Each type of design has advantages and disadvantages. A paired design is often employed to reduce the effect of the natural variation between flasks or replicates. How this may be achieved can be seen by examination of the formula for the unpaired t test [Eq. (3.1)]. A value of t obtained in an unpaired test is the difference between the two treatment means divided by the SE of this difference. If variation among the flasks is large, say from processing them at different times of the day or on different days, it may increase the SE of the difference and lower the value of t even if the difference between the means is large. Notice, however, that in an unpaired design, the t table is entered for 48 DF. Pairing the flasks may reduce the SE because the value of paired t is calculated from the differences between pairs of observations. In other words, the effect of the experimental treatment is being determined within a "matched pair" of flasks sampled at a specific time. Pairing should only be considered, however, if there is good evidence that it actually reduces the variability, for example, pairing supplemented and unsupplemented flasks on the same day when the day of measurement is known to significantly affect the measurement. If there is no reduction in the SE by pairing, that is, it does not matter on which day the samples are measured, then there is a disadvantage of the paired design because the t table is entered with only 24 DF (one less than the number of pairs). Entering the t table with a smaller number of DF means that a larger value of t will be required to demonstrate a significant difference between the means.

3.8 CONCLUSION

The t test is a useful method of comparing two groups when the data approximate to a normal distribution. There are two different types of t tests, depending on whether the data are paired or unpaired. In addition, t tests may be one tail or two tail depending on the exact nature of the H_0.

Statnote 4

WHAT IF THE DATA ARE NOT NORMALLY DISTRIBUTED?

Data transformation.
Skew and kurtosis.
Mode and median.
The Mann–Whitney test.
The Wilcoxon signed-rank test.

4.1 INTRODUCTION

The statistical tests described in the first three statnotes make an important assumption regarding the experimental data. The assumption is that the measured quantity, whether an individual measurement, a group mean, or the difference between the means of two groups, are parametric variables, that is, members of normally distributed populations. When this assumption is met, the z and t distributions can be used to make statistical inferences from the data. In some circumstances, however, a variable may not be normally distributed, and this statnote is concerned with the analysis of nonparametric data involving two groups. Further nonparametric tests will be described in Statnotes 17 and 24.

4.2 HOW TO RECOGNIZE A NORMAL DISTRIBUTION

An investigator may know in advance from previous studies whether or not a variable is likely to be normally distributed. In other circumstances, preliminary data may be collected

Statistical Analysis in Microbiology: Statnotes, Edited by Richard A. Armstrong and Anthony C. Hilton
Copyright © 2010 John Wiley & Sons, Inc.

to specifically test whether the data are normal, a procedure that was described in Statnote 1. In many experimental situations, however, there may be insufficient data available to carry out a test of normality, and to obtain such data may be either too expensive or time consuming. In situations such as these, the following points should be considered. First, many continuous variables in microbiology measured to at least three significant figures have a reasonable chance of being normally distributed. Second, the distribution of sample means taken from a population is more likely to be normal than the individual measurements themselves, and, therefore, inferences about means are less susceptible to the problem of nonnormality (Snedecor & Cochran, 1980). Third, moderate departures from normality do not significantly affect the validity of most parametric tests. Consideration of these points may lead to the conclusion that, despite reservations, the data may not depart radically enough from normality to question the validity of a parametric analysis. There are some circumstances, however, where it is more likely that the data will depart significantly from a normal distribution, and a different approach to the analysis may be required.

4.3   NON-NORMAL DISTRIBUTIONS

The two most common ways in which the distribution of a variable may deviate from normality are called skew and kurtosis. Most statistical software will provide tests of these properties, and tables of significance of the relevant statistics are also available (Snedecor & Cochran, 1980; Fisher & Yates, 1963). It is important to note that some distributions may deviate from normal in more complex ways, and, therefore, the absence of significant skew and kurtosis may not guarantee that a distribution is normal.

A skewed distribution is asymmetrical and the mean is displaced either to the left (*positive skew*) or to the right (*negative skew*). By contrast, distributions that exhibit kurtosis are either more "flat topped" (*negative kurtosis*) or have longer tails than normal (*positive kurtosis*). Figure 4.1 shows the frequency distribution of the bacterial counts

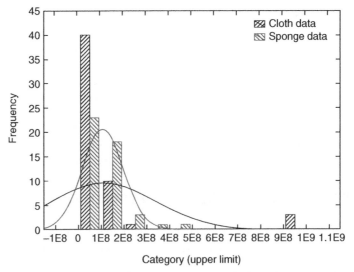

Figure 4.1. Frequency distribution of bacterial counts on cloths and sponges. Fitted curves are those of the normal distribution. The *X* axis is expressed in scientific format.

made on 54 sponges and 46 cloths introduced in Statnote 1. In both cases, the distributions are asymmetrical with the means located to the left of the histogram, and, therefore, they exhibit positive skew. As a result, the arithmetic mean is no longer a good description of the central tendency of such a distribution. There are two additional statistics, however, that can be used to describe the central tendency of a skewed distribution. First, the *mode* is the value of the variable X with the highest frequency, that is, the maximum point of the curve. Second, the *median* is the middle value of X, that is, if all the values of X were listed in ascending or descending order, the median would be the middle value of the array. Little progress has been made in devising statistical tests based on the mode, but there are tests of the differences between the medians of two groups.

An important property of a nonnormal distribution is that the SD is no longer an accurate descriptor of the spread of a distribution with a given mean. Hence, z and t tables cannot be used to predict the proportion of observations that fall a given distance from the mean. On reporting frequency distributions from large samples that are not normally distributed, investigators often quote the *percentiles* of the distribution, for example, the 90th percentile of a distribution is the score such that 90% of the observations fall short of and 10% exceed the score (Snedecor & Cochran, 1980).

4.4 DATA TRANSFORMATION

One method of analyzing nonnormal data is to convert or *transform* the original measurements so that they are expressed on a new scale that is more likely to be normally distributed. Parametric statistics can then be carried out on the transformed values. There are three common circumstances in which such a transformation should be considered. First, if the data are percentages and especially if the majority of the observations are close to 0 or 100%. Percentage data can be transformed to an *angular* or *arcsin* scale defined as follows:

$$\text{Angular measurement} = \sin^{-1}\sqrt{\%/100} \qquad (4.1)$$

Statistical software will often provide this transformation, or it can be obtained from statistical tables [see Table X in Fisher and Yates (1963)]. Percentage data are often skewed when the mean is small or large, and, consequently, the effect of the transformation is that percentages near 0 or 100% are "spread out" to a greater degree than those near the mean so as to increase their variance. A paired or unpaired t test can then be carried out using the transformed values (see Statnote 3). Second, data that comprise small whole numbers or quantities assessed using a score that has a limited scale, for example, the abundance of bacteria may be scored on a scale from 0 (absent) to 5 (abundant), are unlikely to be normally distributed. In this case, a transformation of the data to square roots, that is, $\sqrt{x}$, or $\sqrt{(x+1)}$ if many zeroes are present, may make the scores more normally distributed. Third, the t test described in Statnote 3 also assumes *homogeneity of variance*, that is, that the degree of variability is similar for both groups of observations. It is not unusual, however, for results from a 'control' group to be more consistent, that is, show less variance, than values from an experimentally treated group. In this case, a transformation of the original measurements to a logarithmic scale may equalize the variance and, in addition, may improve the normality of the data. Methods of explicitly testing the assumption of constant variance will be described in Statnote 8.

4.5 SCENARIO

An alternative approach to transformation is to carry out a nonparametric test. As an illustration, we return to the scenario described in Statnote 1. To recapitulate, given the intrinsic structural and compositional differences between cloths and sponges, a study was envisaged to investigate if one material provided a more favorable environment for bacterial survival than the other. A total of 54 "in-use" dishcloths and 46 sponges were collected from domestic kitchens and the aerobic colony count of each determined in the laboratory.

4.6 DATA

The frequency distributions of the counts from both materials are shown in Figure 4.1. In Statnote 1, these distributions were tested for normality, and it was concluded that the cloth data exhibited a significant deviation from normal, whereas the sponge data did not deviate significantly from a normal distribution. However, it may be prudent to conclude that the data as a whole do not conform closely enough to a normal distribution to use the parametric t tests described in Statnote 3. An alternative approach is to use a distribution-free or nonparametric test. These tests can be used regardless of the shape of the underlying distribution as long as the samples being compared can be assumed to come from distributions of the same general type.

4.7 ANALYSIS: MANN–WHITNEY U TEST (FOR UNPAIRED DATA)

4.7.1 How is Analysis Carried Out?

To illustrate this test and to simplify the calculations, we will use data from a sample of 10 cloths and 10 sponges only (see Table 4.1). The Mann–Whitney U test (Mann & Whitney, 1947) can be carried out on two independent groups of data (A, B) and is the nonparametric equivalent of the unpaired t test (see Statnote 3). Although most statistical software will carry out this test, it is still useful to understand its "mechanics," which are shown in Table 4.1. First, ranks 1, 2, 3, … , are assigned to the whole set of observations, regardless of group. A rank of 1 is given to the lowest count, 2 to the next lowest, and so forth. with repeated values, called *ties*, given the mean of the ranks that would have been assigned to those values if they had been different. The ranks of each group are then added together separately to give the totals R_A and R_B. The quantities U_A and U_B are then calculated as shown in Table 4.1. Whichever is the smaller of U_A and U_B is taken to the statistical table of Wilcoxon's U to judge the significance of the difference between cloths and sponges (e.g., Snedecor & Cochran, 1980; Table A10). The lesser U has to be equal to or less than the tabulated value for significance, that is, low values of U indicate a significant difference between the groups. Note that this is the opposite of how many parametric statistical tables work.

4.7.2 Interpretation

In the present example, a value of $U = 16.5$ was obtained that is less than the value tabulated at $P = 0.05$. Hence, there is evidence that the sponges harbor considerably more bacteria than the cloths. For larger samples, outside the range of the statistical table, the

TABLE 4.1   Comparison of Number of Bacteria on 10 Cloths and Sponges (Two Independent Groups, Mann–Whitney U Test)

Cloths (A)		Sponges (B)	
Count	Rank	Count	Rank
1.8×10^6	4	1.1×10^8	13
1.8×10^7	6	2.2×10^8	20
2.0×10^7	7	4.6×10^6	5
5.9×10^7	10	9.8×10^7	11.5
1.6×10^8	19	1.3×10^8	15.5
2.0×10^5	2	1.3×10^8	15.5
9.8×10^7	11.5	1.5×10^8	18
1.1×10^6	3	4.7×10^7	9
6.9×10^4	1	1.4×10^8	17
3.0×10^7	8	1.2×10^8	14

1. Add up the ranks for each group: $R_A = 71.5$, $R_B = 138.5$.
2. Calculate $U_A = [n_A(n_A + 1)/2 + (n_A n_B)] - R_A = 83.5$ where n_A and n_B are the number of observations in each group.
3. Calculate $U_B = [n_B(n_B + 1)/2 + (n_A n_B)] - R_B = 16.5$.
4. The smaller U (in this case 16.5) is the test statistic.
5. Lesser U must be $\leq$ Wilcoxon's tabulated U for significant difference.
6. For larger samples: $Z = (\mu - T - 1/2)\sigma$ where T is the smaller rank sum and $\sigma = \sqrt{n_B \mu / 6}$ and $\mu = n_A(n_A + n_B + 1)/2$.

data may approach a normal distribution more closely and a value of Z can be calculated (Table 4.1), the statistic being referred to the Z table.

4.8   ANALYSIS: WILCOXON SIGNED-RANK TEST (FOR PAIRED DATA)

4.8.1   How Is the Analysis Carried Out?

If the data in the two groups are paired (see Statnote 3), then the appropriate nonparametric test is the Wilcoxon signed-rank test (Wilcoxon, 1945). To illustrate this test (Table 4.2), data on the number of bacteria were collected on a cloth and sponge on each of 10 separate occasions. Hence, we do not have 2 independent samples as in the previous example. In this case, there is a connection between a particular cloth and sponge in that the data for each pair were collected on a specific occasion. Essentially, the data are subtracted for each pair of observations (A – B). Omitting zero differences, ranks (r) are applied to all of the remaining values of A – B, regardless of whether the difference is positive or negative. If ties occur between positive and negative columns, the ranks are amended in any such run of ties to the mean rank within the run as before. The positive and negative signs are restored to the ranks and the positive and negative ranks added up. R is the smaller of the two sums of ranks and is taken to the table of the Wilcoxon signed-rank statistic T to obtain a P value (Snedecor & Cochran, 1980; Table A9). The value of R has to be equal to or less than the value of T in the $P = 0.05$ column to demonstrate a significant difference between the two groups.

TABLE 4.2   Comparison of Bacteria on Pairs of Cloths and Sponges Sampled on 10 Occasions (Two Dependent Groups, Wilcoxon Signed-Rank Test)

Occasion	Cloth (A)	Sponge (B)	$A - B$	Rank
1	1×10^4	4.6×10^6	-4.5×10^6	-2
2	3.3×10^7	9.8×10^7	-6.5×10^7	-6
3	5.7×10^7	1.3×10^8	-7.3×10^7	-7
4	1.9×10^7	1.3×10^8	1.11×10^8	-9
5	1.2×10^4	6.0×10^2	$+1.1 \times 10^4$	$+1$
6	8.8×10^2	4.7×10^7	-4.7×10^7	-5
7	2.6×10^6	1.4×10^8	-1.14×10^7	-3
8	3.3×10^7	1.2×10^8	-8.7×10^7	-8
9	8.7×10^6	2.1×10^8	-2.0×10^8	-10
10	7.6×10^7	$1.1 \times 10_8$	-3.4×10^7	-4

1. Subtract each pair of counts $A - B$.
2. Assign ranks (r) to differences, ignoring the sign of the difference.
3. Restore the signs and add up the positive and negative ranks.
4. Compare the lesser R (in this case $+R = 1$) with the tabulated Wilcoxon's signed-rank statistic T, $R \leq T$ for significance.
5. For larger samples $Z = (\mu - T - 1/2)\sigma$ where T is the smaller rank sum and $\sigma = \sqrt{(2n+1)\mu/6}$ where n = number of pairs and $\mu = n(n + 1)/4$.

4.8.2   Interpretation

In this case, our value of $R = 1$ was less than the tabulated value, indicating that sponges harbor more bacteria than the cloths. With larger numbers of observations, a value of Z can be calculated and referred to tables of the normal distribution.

4.9   COMPARISON OF PARAMETRIC AND NONPARAMETRIC TESTS

What is the relative sensitivity of a parametric compared with a nonparametric test and what happens if they are used incorrectly? If a t test is used on nonnormal data, the significance probabilities are changed, and the sensitivity or "power" of the test is altered, and this can result in erroneous conclusions especially if treatment effects are of borderline significance. With nonparametric tests, the significance levels remain the same for any continuous distribution with the exception that they are affected by the number of zeroes and tied values in the Wilcoxon signed-rank test (Snedecor & Cochran, 1980). With large normally distributed samples, the efficiency of a nonparametric test is about 95% compared with the t test. With nonnormal data from a continuous distribution, however, the efficiency of a nonparametric test relative to t never falls below 86% in large samples and may be greater than 100% for distributions that are highly skewed.

4.10   CONCLUSION

When testing the difference between two groups and if previous data indicate nonnormality, then the data can either be transformed if they comprise percentages, integers, or scores or alternatively can be analyzed using a nonparametric test. If there is uncertainty whether the data are normally distributed, deviations from normality are likely to be small if the data are measurements made to at least three significant figures. Unless there is clear evidence that the distribution is nonnormal, it is often more efficient to use a parametric test. It is poor statistical practice to carry out both parametric and nonparametric tests on the same set of data and then choose the result most convenient to the investigator!

Statnote 5

CHI-SQUARE CONTINGENCY TABLES

The 2×2 contingency table.

Chi-square (χ^2) distribution.

Yate's correction.

Fisher's 2×2 exact test.

The $R \times C$ contingency table.

5.1 INTRODUCTION

Statnotes 2 and 3 describe the application of statistical methods to measurement data. Measurement data are expressed in units; they are continuous variables, and, in many cases, fulfill the requirements of the normal distribution. In some studies, however, the data are not measurements but comprise counts or frequencies of particular events. Such data are often analyzed using the chi-square (χ^2) distribution. An example of the use of this statistic to test whether an observed distribution of frequencies came from a normal distribution (goodness-of-fit test) was described in statnote 1. The objective of this Statnote is to extend these methods to the analysis of frequencies classified according to two different variables.

Statistical Analysis in Microbiology: Statnotes, Edited by Richard A. Armstrong and Anthony C. Hilton
Copyright © 2010 John Wiley & Sons, Inc.

5.2 SCENARIO

Methicillin-resistant *Staphlococcus aureus* (MRSA) is a significant cause of infection as a result of treatment in a hospital as well as community morbidity and mortality. Hence, over the past two decades, MRSA has become a worldwide problem exacerbated by the emergence of drug-resistant isolates. Such isolates demonstrate a reduced susceptibility to almost all clinically available antibiotics. It is generally accepted that sublethal exposure of bacteria to antibiotics can promote the rapid development of resistance and that this situation may be more likely to occur in a hospital setting than in the community. It might be hypothesized, therefore, that isolates of MRSA from a hospital (HA-MRSA) would demonstrate an enhanced resistance profile to antibiotics compared to MRSA isolated from the community (CA-MRSA).

To test this hypothesis, 197 isolates of MRSA consisting of 95 HA-MRSA and 102 CA-MRSA were isolated from soft tissue infections and screened for their sensitivity to a panel of 10 antibiotics using the British Society for Antimicrobial Chemotherapy (BSAC) disk diffusion method. Isolates were designated as resistant (*R*) or sensitive (*S*). If the hospital is providing an environment that promotes the development of antibiotic resistance, then it might be expected that HA-MRSA would demonstrate a greater than average spectrum of resistance (i.e., to ≥5 antibiotics of the 10 screened) than those isolated from the community. The potential significance of the association between antibiotic sensitivity and location can be investigated using a χ^2 test.

5.3 DATA

The data comprise the frequencies of hospital-acquired and community-acquired MRSA antibiotic sensitivities, which exhibit resistance to five or more antibiotics and to less than five antibiotics. The data are tabulated in the form of a 2×2 contingency table and are presented in Table 5.1.

TABLE 5.1 Association between Hospital-Acquired (HA) and Community-Acquired (CA) MRSA Antibiotic Sensitivities when $N > 20$

MRSA Isolate	2×2 Contingency Table		
	Resistant to ≥5 Antibiotics	Resistant to <5 Antibiotics	Total
HA-MRSA	42	53	95
CA-MRSA	5	97	102
Total	47	150	$197 = N$

1. The expected frequency (E_F) in each cell is calculated as (Row total × Column total)/N.
2. Hence, the expected frequency of HA-MRSA isolates resistant to ≥5 antibiotics is $(95 \times 47)/197 = 22.66$. This calculation is repeated for each of the four cells of the table.
3. Calculate $\chi^2 = \Sigma(O_F - E_F)^2/E_F$. In this cases $\chi^2 = 41.84$ (39.70 with Yate's correction) ($P < 0.001$) with 1 DF.

5.4 ANALYSIS: 2 × 2 CONTINGENCY TABLE

5.4.1 How Is the Analysis Carried Out?

In Table 5.1, 44% of the HA-MRSA isolates were resistant to ≥ 5 antibiotics as against 4.9% of the CA-MRSA isolates. Is this difference sufficient to conclude that there is an association between the antibiotic sensitivity profile of the isolate and its location? First, the expected frequencies (F_e) are calculated for each cell of the 2 × 2 table and subtracted from the observed frequencies (F_o). χ^2 is the sum of the squares (SS) of these deviations divided by the appropriate expected frequency:

$$\chi^2 = \sum \frac{(F_o - F_e)^2}{F_e} \tag{5.1}$$

The value of χ^2 is taken to the χ^2 table for 1 DF to obtain the probability that the value of the statistic would occur by chance if there were no differences between the degree of resistance at the two locations.

5.4.2 Interpretation

The calculated value of χ^2 ($\chi^2 = 41.84$) is considerably greater than the value tabulated at the 5% level of probability for 1 DF. This is a value that would occur rarely by chance—in fact, less than 1 in a 1000—and hence we conclude that there is an association between the antibiotic sensitivity profile of an isolate and its location. Caution is necessary when interpreting the results of χ^2 tests in observational studies (Snedecor & Cochran, 1980). There may be many factors that vary between a hospital and community setting that could influence the antibiotic resistance profile of a MRSA strain, some of which may be wholly or partly responsible for an observed significant difference.

There are four frequencies in a 2 × 2 table and yet the test only has 1 DF. To understand why a 2 × 2 table has only 1 DF, it is instructive to examine the deviations of the observed from the expected frequencies for each cell of the table individually. Examination of these deviations will show that they are all the same apart from their sign, that is, in a 2 × 2 table there is only a single "independent" estimate of the deviation of the observed from the expected frequency. An additional statistic that is sometimes given by statistical software is called ϕ^2 (phi square) and is a measure of the degree of *correlation* between the two variables in a 2 × 2 table.

5.4.3 Yate' Correction

Statistical software usually includes the option of calculating χ^2 with Yate' correction. This correction improves the estimate of χ^2 in a 2 × 2 table when the frequencies are small (say <10). The absolute value of the difference between the observed and expected frequencies is reduced by 0.5 before squaring. The effect of this is to make the estimate of χ^2 slightly more conservative when the table contains small frequencies. Yate' correction applied to the above example gives a value of $\chi^2 = 39.70$.

5.5 ANALYSIS: FISHER'S 2 × 2 EXACT TEST

The χ^2 test described above is only an approximate test when applied to a 2 × 2 table, and the approximation becomes poorer as sample size decreases. One remedy is to apply

TABLE 5.2 Association between Hospital-Acquired (HA) and Community-Acquired (CA) MRSA Antibiotic Sensitivities when $N < 20$

| MRSA Isolate | 2×2 Contingency Table | | |
	Resistant to ≥5 Antibiotics	Resistant to <5 Antibiotics	Total
HA-MRSA	A	B	(A + B)
CA-MRSA	C	D	(C + D)
Total	(A + C)	(B + D)	Total = N

1. If N is less than 12, use Fisher's 2×2 exact test and calculate the probability (P) of this particular outcome among all possible outcomes with the same row and column totals:

$$P = \frac{A! \times B! \times C! \times D!}{CA! \times CB! \times DA! \times DB! \times N!}$$

2. If N is larger than 12, then a calculation based on logarithms can be used (see Fisher and Yates Table 30; Dawkins 1975).

Yate' correction as described above. An alternative to χ^2, called Fisher's 2×2 exact test (Fisher, 1922), can also be used and the principle is illustrated in Table 5.2. This test should be applied if the total sample size is less than 20 or if N lies between 20 and 40 and the expected frequency is less than 5. When the total of the observations is small, say less than 12, the probability of a particular distribution of values in a 2×2 table being obtained, given the particular row and column totals, can be calculated directly from the data. If the total is larger than 12, then a more complex calculation can be made using logarithms (Dawkins, 1975).

5.6 ANALYSIS: ROWS × COLUMNS ($R \times C$) CONTINGENCY TABLES

It is possible to analyze two variables with more than two rows and columns, and this is termed a rows (R) × columns (C) contingency table. For example, antibiotic resistance may have been tested for three or more strains or locations simultaneously. To make the test, the expected frequency is calculated for each cell of the table as in Table 5.1. The value of χ^2 is then calculated using the usual formula, and the value of χ^2 compared with the χ^2 distribution, entering the table for (Number of rows − 1) × (Number of columns − 1) DF (Snedecor & Cochran, 1980). If a significant χ^2 is obtained, the $R \times C$ table may need to be broken down into smaller tables to compare some of the isolates in more detail.

5.7 CONCLUSION

When the data are counts or the frequencies of particular events and can be expressed as a contingency table, then they can be analyzed using the χ^2 distribution. When applied to a 2×2 table, the test is approximate and care needs to be taken in analyzing data when the expected frequencies are small, either by applying Yate's correction or by using Fisher's 2×2 exact test. Larger contingency tables can also be analyzed using this method. Note that it is a serious statistical error to use any of these tests on measurement data, that is, data expressed in units.

Statnote 6

ONE-WAY ANALYSIS OF VARIANCE (ANOVA)

Introduction to ANOVA.

The one-way ANOVA in a randomized design.

Assumptions of ANOVA.

6.1 INTRODUCTION

Studies in applied microbiology often involve comparing either several different groups or the influence of two or more factors at the same time (Armstrong & Hilton, 2004). For example, an investigator may be interested in the degree of microbial contamination on coins collected from a number of different commercial properties, and the data analysis would involve a comparison of microbial numbers from the different locations. In addition, an investigator may wish to compare the pattern of transfer of bacteria from different dishcloths, either rinsed or not, on to a food preparation surface (Hilton & Austin, 2000). In this case, two factors influence microbial numbers, namely, type of dishcloth and rinsing treatment. An investigator may wish to establish whether each factor influenced microbial numbers individually and, in addition, whether there was an interaction between them, that is, does rinsing have the same effect on numbers of transferred bacteria when different dishcloths are used? The most appropriate method of statistical analysis of this type of experiment is analysis of variance (ANOVA).

Analysis of variance is the most effective method available for analyzing more complex data sets. It is, however, a method that comprises many different variations, each

Statistical Analysis in Microbiology: Statnotes, Edited by Richard A. Armstrong and Anthony C. Hilton
Copyright © 2010 John Wiley & Sons, Inc.

of which applies in a particular experimental context (Armstrong et al., 2000, 2002). Hence, it is possible to apply the wrong type of ANOVA and to draw erroneous conclusions from the results. As a consequence, various different experimental designs and their appropriate ANOVA will be discussed in these statnotes. This statnote is an introduction to ANOVA and describes how ANOVA came to be invented, the logic on which it is based, and the basic assumptions necessary for its correct use. The application of the method to the simplest type of scenario, namely a one-way ANOVA in a completely randomized design is then described.

6.2 SCENARIO

An experiment was set up to measure the degree of bacterial contamination on 2p (pence) coins collected from three types of business premises, namely a butcher shop, a sandwich shop, and a newsstand. A sample of four coins was collected at random from each property. The number of bacterial colonies present on each coin was then estimated by dilution plating techniques.

6.3 DATA

The data comprise four randomly obtained measurements of the bacterial contamination classified into three groups (the property) and therefore comprise a *one-way classification in a randomized design*. The data are presented in Table 6.1.

TABLE 6.1   Number of Bacteria Isolated from 2p Coins Collected from Three Types of Property[a]

	Butcher Shop	Sandwich Shop	Newsstand
	140	2	40
	108	21	5
	76	0	5
	400	42	0
Mean	181	16	13
SE	74.16	9.80	9.24

1. Total sum of squares (SS) = $\Sigma(x_{ij} - X^*)^2$ = 142238.917.
2. Between groups (property) SS = $\Sigma(\Sigma T_i - H^*)^2$ = 74065.167.
3. Error SS = Total SS − Between groups SS = 68173.75.

ANOVA table					
Source	DF	SS	MS	F	P
Between property	2	74065.167	37032.583	4.89	0.037
Error	9	68173.75	7574.861		

[a] SE = Standard error of the mean, DF = degrees of freedom, SS = sums of squares, x_{ij} = individual counts, X^* = overall mean of all 12 bacterial counts, T_i = column total, H^* = mean of three column totals, MS = mean square, F = variance ratio, P = probability.

6.4 ANALYSIS

6.4.1 Logic of ANOVA

If only two types of property were present, then the H_0 that there was no significant mean difference in the numbers of bacteria at the two locations could be tested using Student's t test (see Statnote 3). The statistic t is the ratio of the difference between the means of two properties to a measurement of the variation between the individual coins pooled from both properties. A significant value of Student's t indicates that the H_0 should be rejected and that there was a real difference in bacterial numbers between the two properties.

This method of analysis could be extended to three or more different locations. To compare all pairs of properties, however, three t tests would be necessary (Fig. 6.1). There is a problem in making multiple comparisons between these means, however, because not all of the comparisons can be made independently. For example, if bacterial numbers on coins from the butcher shop were significantly greater than those from the sandwich shop but similar to those from the newsstand, it follows that numbers in the newsstand "should" be greater than those in the sandwich shop (Fig. 6.1). However, the latter comparison would not have been tested independently but is essentially "predetermined" once the first two comparisons have been established. To overcome this problem, ANOVA was developed by Sir Ronald Fisher in the 1920s and provides a single statistical test of the H_0 that the means of the bacterial numbers from the three properties are identical (Fisher, 1925, 1935).

6.4.2 How Is the Analysis Carried Out?

The essential feature of an ANOVA is that the total variation between the observations is calculated and then broken down ("partitioned") into portions associated with, first, differences between the three properties and, second, variation between the replicate coins sampled within a property. The calculations involved are shown in Table 6.1. The sum of squares (SS) of the deviations of the observations (x_{ij}) from their mean (X^*) is used as a measure of the total variation of the data while the SS of the three property group means

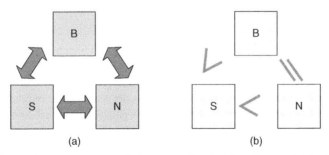

(a)   (b)

Figure 6.1. Comparisons between multiple groups (B = butcher, S = sandwich shop, N = newsstands: (a) Three t tests would be necessary to compare the three properties. (b) If bacterial numbers on coins from the butcher's shop were significantly greater than those from the sandwich shop but similar to those from the newsstand, it follows that numbers in the newsstand should be greater than those in the sandwich shop.

from their "overall" mean is a measure of the treatment effect and is calculated from the column totals (T_i). In addition, there is variation between the four replicate coins within each property. This variation is often called the *residual* or *error* variation because it describes the natural variation that always exists between replicates. Variation between replicates within a group is calculated as the SS of the raw data (x_{ij}) in each column from their column mean. The SS calculated from each column are then added together to give the error SS. In this case, however, there are only two sources of variation present, that is, between the three properties and error and the error SS can be calculated more directly by subtraction.

If there are no significant differences between the means of the 3 properties, the 12 observations are distributed about a common population mean μ. As a result, the variance (also called the mean square) calculated from the between-treatments SS and the error SS should be estimates of the same quantity. Testing the difference between these two mean squares is the basis of an ANOVA. The statistics are set out in an ANOVA table (Table 6.1). To compare the between-treatments and error mean squares, the SS are divided by their appropriate DF. We can also regard the DF of a quantity as the number of observations minus the number of parameters estimated from the data required to calculate the quantity (see also Statnote 2). Hence, the total and between-property SS each have 11 and 2 DF, respectively, one less than the total number of observations or groups. This is because the mean of the x_{ij} values and the mean of the 3 treatment totals were calculated from the data to obtain the SS. The error sum of squares has 9 DF because the column means are used in the calculation, that is, there are 3 DF in each of the 3 columns ($4 - 1 = 3$), making 9 in total.

The between-property mean square is then divided by the error mean square to obtain the variance ratio. This statistic was named F (hence, F test) in honor of Fisher by G.W. Snedecor (Snedecor & Cochran, 1980). The value of F indicates the number of times the between-property mean square exceeds that of the error mean square. This value is compared with the statistical distribution of the F ratio to determine the probability of obtaining a statistic of this magnitude by chance, that is, from data with no significant differences between the group means. If the value of F is equal to or greater than the value tabulated at the 5% level of probability, then the H_0 that the three treatment means are identical is rejected.

6.4.3   Interpretation

In the present scenario, a value of $F = 4.89$ was obtained, which has a P value of 0.0365, that is, there is less than a 5% chance of obtaining an F ratio of this magnitude by chance alone. This result indicates a real difference between the bacterial counts from the three properties. Note that the analysis relates only to the three individual properties studied. It would not be possible to make a more general statement about property of this type from these data. This would require a random sample of each property type to be sampled so that an estimate could be obtained of the variation between similar types of property.

The F test of the group means is only the first stage of the data analysis. The next step involves a more detailed examination of the differences between the means. A variety of methods are available for making such tests, often referred to as post hoc tests, and several of these tests are usually available in statistical packages and will be discussed in the next statnote.

6.5   ASSUMPTIONS OF ANOVA

Analysis of Variance makes certain assumptions about the nature of the experimental data that have to be at least approximately true before the method can be validly applied. An observed value in Table 6.1 (x_{ij}) can be considered to be the sum of three parts: (1) the overall mean of the observations (μ), (2) a treatment or class deviation, and (3) a random element drawn from a normally distributed population. The random element reflects the combined effects of natural variation between replications and errors of measurement. ANOVA assumes, first, that these errors are normally distributed with a zero mean and standard deviation s, second, that although the means may vary from group to group, the variance is constant in all groups, and, third, that effects of individual treatments are additive rather than multiplicative.

Failure of one or more of these assumptions affects both the significance levels and the sensitivity of the F tests. Experiments are usually too small to test whether these assumptions are actually true. In many biological and medical applications, in which a quantity is being measured, however, the assumptions are likely to hold well (Cochran and Cox, 1957; Ridgman, 1975). In many applications in microbiology in which bacterial numbers are being estimated, the assumptions may not hold. There are often two problems when the data comprise numbers of microbes. First, small whole numbers, especially if there are many zeros, are unlikely to be normally distributed, and, second, a large range of bacterial numbers may be present resulting in heterogeneous variances within the different groups. The latter problem is evident in the example analyzed in Table 6.1 in which the SE for the three properties varies markedly. If there is doubt about the validity of the assumptions, significance levels and confidence limits must be considered to be approximate rather than exact. If the data are small whole numbers, then the assumptions are unlikely to be met. If the assumptions are not met, then transformation of the data will often allow an ANOVA to be carried out. For example, in Table 6.1, a transformation of the data to logarithms could have equalized the variances for the three properties (see Statnote 4). Alternatively, there are nonparametric forms of ANOVA that could be carried out in some circumstances and these will be discussed in Statnote 24.

6.6   CONCLUSION

If the data comprise three or more groups, then the t test should not be used to make comparisons between pairs of group means, ANOVA being the most appropriate analysis. When there are three or more groups and the replicate measurements are collected randomly for each group, the experimental design is often described as a one-way classification in a randomized design. There are, however, many other experimental designs, and each has its appropriate ANOVA. The most frequently encountered of these designs will be described in the following statnotes.

Statnote 7

POST HOC TESTS

Comparison of group means.

Planned comparisons between the means.

Post hoc tests including Fisher's PLSD, Scheffé's test, Dunnett's test, and Bonferroni's test.

7.1 INTRODUCTION

In Statnote 6, the application of analysis of variance (ANOVA) to the analysis of three or more independent groups (the one-way ANOVA in a randomized design) was described. The first stage of the analysis was to carry out a variance ratio test (F test) to determine whether all the group means could be considered to come from the same population. If treatment groups are few, say three or four, a nonsignificant F test suggests that it is unlikely that there are meaningful differences among the group means and no further analysis would be required. A significant value of F, however, suggests real differences among the group means, and the next stage of the analysis would involve a more detailed examination of these differences. There are various options available for the subsequent analysis of the data, depending on the objectives of the experiment. Specific comparisons may have been planned before the experiment was carried out, decided during the analysis stage, or comparisons between all possible combinations of the group means may have been envisaged. This statnote discusses the various methods available for a more detailed analysis of the data following a one-way ANOVA.

Statistical Analysis in Microbiology: Statnotes, Edited by Richard A. Armstrong and Anthony C. Hilton
Copyright © 2010 John Wiley & Sons, Inc.

7.2 SCENARIO

An experiment was designed to investigate the efficacy of two commercial plasmid prep kits compared to a standard alkaline–SDS (sodium dodecyl sulfate) lysis protocol. A 5-ml overnight recombinant *Escherichia coli* culture containing a high copy plasmid was harvested by centrifugation and the pellet resuspended in 100 μl of lysis buffer. Plasmid DNA (deoxyribonucleic acid) was subsequently extracted from the cell suspension using a standard SDS lysis protocol or a commercially available kit following the manufacturer's instructions. In total, 10 samples were allocated at random and processed using each of the three extraction methods under investigation. Following extraction, the purified plasmid DNA pellet was dissolved in 50 μl of water and the concentration determined spectrophotometrically at 260 nm.

7.3 DATA

The data comprise the yields of plasmid DNA using the three different preparation methods with 10 replications of each method and are presented in Table 7.1. Hence, as in Statnote 6, the data comprise a single classification in a randomized design, that is, the data are classified in only one way, according to preparation method.

7.4 ANALYSIS: PLANNED COMPARISONS BETWEEN THE MEANS

An experiment may have been designed to test specific (*planned*) comparisons between the treatment means. Planned comparisons are hypotheses specified before the experiment has been carried out, whereas post hoc tests are for further explanation after a significant effect has been found.

7.4.1 Orthogonal Contrasts

The basic strategy for planned comparisons is to partition the total between-group variation (the SS between groups) among the various hypotheses, which are called "contrasts," and which are then analyzed separately either by an *F* test or a *t* test. If this procedure was carried out for all possible comparisons between the means, then the SS for all contrasts would be greater than the treatment SS as a whole since the comparisons overlap and are based on the same sources of variance. Strictly speaking, such comparisons cannot be made independently of each other. As a result, comparisons must be constructed so that they are not overlapping, that is, they have to be "orthogonal." Essentially, orthogonal comparisons have no common variance and their coefficients sum to zero. Hence, the SS can be calculated for each contrast and a test of significance made on each. The number of possible contrasts is equivalent to the number of DF of the treatment groups in the experiment. Hence, if an experiment employs three groups, as in the present scenario, then only two contrasts can be validly tested, that is, there are two orthogonal comparisons between the treatment means. This approach has two advantages. First, there is no problem as to the validity of the individual comparisons, a problem that is present to some extent with all conventional post hoc tests. Second, the comparisons provide direct tests of the hypotheses of interest. Most commercially available software will allow for valid contrasts to be tested for a range of experimental designs.

TABLE 7.1 Two Commercial Plasmid Prep Kits (A, B) (Plasmid Yield in mg) Compared to Standard Alkaline–SDS Lysis Protocol Using Planned Comparisons and post hoc Tests

Alkaline–SDS Lysis	Commercial Kit A	Commercial Kit B
1.7	3.1	4.7
2	2.2	3.5
1.2	2.8	2.6
0.5	4.8	4.3
.9	5	3.8
1	1.9	4.5
1.4	2	4
2.7	3.6	1.9
3.2	4.1	2.8
0.7	4.7	4.6

ANOVA

Source of Variation	Sums of Squares	DF	Mean Square	F
Treatments	27.3807	2	13.690	13.28 ($P < 0.001$)
Error	27.998	27	1.0370	

Planned Comparisons

Contrast	Estimate	Standard Error (SE)	t
1. Standard vs. (Kit A + Kit B)/2	4.03	0.79	5.109 ($P < 0.001$)
2. Kit A vs. Kit B	0.25	0.45	0.54 ($P > 0.05$)

Post hoc Tests

	Standard vs. Kit A	Standard vs. Kit B	Kit A vs. Kit B
Fisher PLSD	$P < 0.001$	$P < .001$	$P > 0.05$
Tukey–Kramer HSD	$P < 0.001$	$P < 0.001$	$P > 0.05$
SNK	$P < 0.001$	$P < 0.001$	$P > 0.05$
Scheffé	$P < 0.001$	$P < 0.001$	$P > 0.05$

7.4.2 Interpretation

An example of this approach is shown in Table 7.1. In our scenario, we compared two commercial plasmid prep kits with a standard alkaline–SDS lysis protocol. Two valid contrasts are possible using this experimental design. First, the mean of the results from the two commercial prep kits can be compared with the standard method, namely do the commercial kits on average improve plasmid yield (contrast 1)? Second, the two commercial prep kits themselves can be compared (contrast 2). Contrast 1 is highly significant ($t = 5.11$, $P < 0.001$), indicating the superiority of the commercial kits over the standard method, but contrast 2 is not significant ($t = 0.54$, $P > 0.05$), showing that the two commercial kits did not differ in their efficacy.

7.5 ANALYSIS: POST HOC TESTS

7.5.1 Common Post Hoc Tests

There may be circumstances in which multiple comparisons between the treatment means may be required but no specific hypotheses were identified in advance. There are a variety of methods available for making post hoc tests. The most common tests included in many commercially available software packages are listed in Table 7.2 (Abacus Concepts, 1993; Armstrong et al., 2000). These tests determine the critical differences that have to be exceeded by a pair of group means to be significant. However, the individual tests vary in how effectively they address a particular statistical problem and their sensitivity to violations of the assumptions of ANOVA. The most critical problem is the possibility of making a type 1 error, that is, rejecting the H_0 when it is true (i.e., asserting the presence of an effect when it is absent). By contrast, a type 2 error is accepting the H_0 when a real difference is present. The post hoc tests listed in Table 7.2 give varying degrees of protection against making a type 1 error.

Fisher's protected least significant difference (Fisher's PLSD) is regarded as the most "liberal" of the methods and, therefore, the most likely to result in a type 1 error. All possible pairs of means are evaluated, and the method uses the distribution of t to determine the critical value to be exceeded for any pair of means based on the maximum number of steps between the smallest and largest mean. The Tukey–Kramer honestly

TABLE 7.2 Methods of Making post hoc Multiple Comparisons between Means[a]

Method	N	F	Var	Nm	Use	Error Control
Fisher PLSD	Yes	Yes	Yes	Yes	All pairwise comparisons	Most sensitive to type 1
Tukey—Kramer (HSD)	No	No	Yes	Yes	All pairwise comparisons	Less sensitive to type 1
Spjotvoll–Stoline	No	No	Yes	Yes	All pairwise comparisons	Similar to Tukey–Kramer
Student–Newman– Keuls (SNK)	Yes	Yes	Yes	Yes	All pairwise comparisons	Sensitive to type 2
Tukey– Compromise	No	No	Yes	Yes	All pairwise comparisons	Average of Tukey and SNK
Duncan's multiple range test	No	No	Yes	Yes	All pairwise comparisons	More sensitive to type 1 than SNK
Scheffé's test	No	Yes	No	No	All pairwise comparisons	Most conservative
Games/Howell	No	Yes	No	No	All pairwise comparisons	Conservative
Dunnett's test	No	No	No	Yes	Compare with control	Conservative
Bonferroni	Yes	No	Yes	Yes	All pairwise comparisons and with control	Conservative

[a] PLSD = Protected least significant difference, HSD = honestly significant difference. Column 2 indicates whether equal numbers of replicates (N) in each treatment group are required or whether the method can be applied to cases with unequal N. Column 3 indicates whether a significant overall between treatments F ratio is required, and columns 4 and 5 whether the method assumes equal variances (Var) in the different groups and normality of errors (Nm), respectively. The final column indicates the degree of protection against making a type 1 or type 2 error.

significant difference (Tukey–Kramer HSD) is similar to the Fisher PLSD but is less liable to result in a type 1 error (Keselman & Rogan, 1978). In addition, the method uses the more conservative "Studentized range" rather than Student's t to determine a single critical value that all comparisons must exceed for significance. This method can be used for experiments that have equal numbers of observations (N) in each group or in cases where N varies significantly between groups. With modest variations in N, the Spjotvoll–Stoline modification of the above method can be used (Spjotvoll & Stoline, 1973; Dunnett, 1980a). The Student–Newman–Keuls (SNK) method makes all pairwise comparisons of the means ordered from the smallest to the largest using a stepwise procedure. First, the means furthest apart, that is, a steps apart in the range, are tested. If this mean difference is significant, the means $a - 2$, $a - 3$, and so on, steps apart are tested until a test produces a nonsignificant mean difference, after which the analysis is terminated. The SNK test is more vulnerable to a type 2 rather than a type 1 error.

By contrast, the Tukey "compromise" method employs the average of the HSD and SNK critical values. Duncan's multiple range test is very similar to the SNK method but is more liberal than SNK, the probability of making a type 1 error increasing with the number of means analyzed. One of the most popular methods is Scheffé's S test (Scheffé, 1959). This method allows all pairs of means to be compared and is a very robust procedure to violations of the assumptions associated with ANOVA (see Statnote 6). It is also the most conservative of the methods available, giving maximum protection against making a type 1 error. The Games–Howell method (Games & Howell, 1976) is one of the most robust of the newer methods. It can be used in circumstances where N varies between groups, with heterogeneous variances (see Statnote 6), and when a normal distribution cannot be assumed. This method defines a different critical value for each pairwise comparison, and this is determined by the variances and numbers of observations in each group under comparison. Dunnett's test is used when several group means are each compared to a control mean. Equal or unequal N can be analyzed, and the method is not sensitive to heterogenous variances (Dunnett, 1980b). An alternative to this test is the Bonferroni/Dunn method, which can also be employed to test multiple comparisons between treatment means, especially when a large number of treatments is present.

7.5.2   Which Test to Use?

In many circumstances, different post hoc tests may lead to the same conclusions and which test is actually used is often a matter of fashion or personal taste. Nevertheless, each test addresses the statistical problems in a unique way. One method of deciding which test to use is to consider the purpose of the experimental investigation. If the purpose is to decide which of a group of treatments is likely to have an effect and it is the intention to investigate such effects in more detail, then it is better to use a more liberal test such as Fisher's PLSD. In this scenario, it is better not to miss a possible effect. By contrast, if the objective is to be as certain as possible that a particular treatment does have an effect, then a more conservative test such as the Scheffé test would be appropriate. Tukey's HSD and the compromise method fall between the two extremes and the SNK method is also a good choice. We would also recommend the use of Dunnett's method when several treatments are being compared with a control mean. However, none of these methods is an effective substitute for an experiment designed specifically to make planned comparisons between the treatment means (Ridgman, 1975).

7.5.3   Interpretation

As an example, the data in Table 7.1 were analyzed using four different post hoc tests, namely Fishers PLSD, Tukey–Kramer HSD, the SNK procedure, and Scheffé's test. All four tests lead to the same conclusion, that is, both commercial kits are superior to the standard method, but there is no difference between commercial kits A and B, thus confirming the results of the planned comparisons. Nevertheless, it is not uncommon for various post hoc tests applied to the same data to give different results.

7.6   CONCLUSION

If data are analyzed using ANOVA, and a significant F value is obtained, a more detailed analysis of the differences between the treatment means will be required. The best option is to plan specific comparisons among the treatment means before the experiment is carried out and to test them using contrasts. In some circumstances, post hoc tests may be necessary, and experimenters should choose carefully which of the many tests available to use. Different tests may lead to different conclusions.

Statnote 8

IS ONE SET OF DATA MORE VARIABLE THAN ANOTHER?

Comparing two variances: The two-tail F *test.*

Comparing three or more variances: Bartlett's test, Levene's test, and Brown–Forsythe test.

8.1 INTRODUCTION

It may be necessary to test whether the variability of two or more sets of data differ. For example, an investigator may wish to determine whether a new treatment reduces the variability in the response of a microbial culture to an antibiotic compared with an older more conventional treatment. In addition, an important assumption for the use of the t test (see Statnote 3) or one-way analysis of variance (ANOVA) (see Statnote 6) is that the variability between the replicates that comprise the different groups is similar, that is, that they exhibit *homogeneity of variance*. Replicate measurements within a control and a treated group, however, often exhibit different degrees of variability, and the assumption of homogeneity of variance may need to be explicitly tested. This statnote describes four such tests, namely the variance ratio (F) test, Bartlett's test, Levene's test, and Brown and Forsythe's test.

Statistical Analysis in Microbiology: Statnotes, Edited by Richard A. Armstrong and Anthony C. Hilton
Copyright © 2010 John Wiley & Sons, Inc.

8.2 SCENARIO

We return to the scenario first described in Statnote 3 but with an additional supplement added to provide three groups of data. Hence, an experiment was carried out to investigate the efficacy of two novel media supplements (S_1 and S_2) in promoting the development of cell biomass. Three 10-liter fermentation vessels were sterilized and filled with identical growth media, with the exception that the media in two of the vessels was supplemented with 10 ml of either medium supplement S_1 or S_2. The vessels were allowed to equilibrate and were subject to identical environmental/incubation conditions. The vessels were then inoculated with a culture of the same bacterium at an equal culture density and the fermentation allowed to proceed until all the available nutrients had been exhausted and bacterial growth had ceased. The entire volume of the culture media in each fermentation vessel was then removed and filtered to recover the bacterial biomass, which was subsequently dried and the dry weight of cells measured.

8.3 DATA

The data comprise three independent groups or treatments (a control and two supplements), the experiment being repeated or replicated 25 times. The dry weights of biomass produced in each groups are presented in Table 8.1.

8.4 ANALYSIS OF TWO GROUPS: VARIANCE RATIO TEST

8.4.1 How Is the Analysis Carried Out?

If there were only two groups involved, then their variances could be compared directly using a two-tail variance ratio test (F test) (Snedecor & Cochran, 1980). The larger

TABLE 8.1 Dry Weight of Bacterial Biomass Under Unsupplemented (US) and Two Supplemented (S) Growth Conditions (S_1 and S_2) in a Sample of 25 Fermentation Vessels

US	S_1	S_2	US	S_1	S_2	US	S_1	S_2
461	562	354	506	607	556	518	617	714
472	573	359	502	600	578	527	622	721
473	574	369	501	603	604	524	626	722
481	581	403	505	605	623	529	628	735
482	582	425	508	607	644	537	631	754
482	586	476	500	609	668	535	637	759
494	591	511	513	611	678	542	645	765
493	592	513	512	611	698			
495	592	534	511	615	703			

1. Calculate variances:
   US = 463.36

   S_1 = 447.88

   S_2 = 18695.24
2. Variance ratio test comparing US and S_1: F = 463.36/447.88 = 1.03 (2-tail distribution of F, $P > 0.05$).

variance is divided by the smaller and the resulting F ratio compared with the value in a statistical table of F to obtain a value of P, entering the table for the number of DF of the numerator and denominator. This test uses the two-tail probabilities of F because it is whether or not the two variances differ that is being tested rather than whether variance A is greater than variance B. Hence, this calculation differs from that carried out during a typical ANOVA, since in the latter. it is whether the treatment variance is larger than the error variance that is being tested (see Statnote 6). Published statistical tables of the F ratio (Fisher & Yates, 1963; Snedecor & Cochran, 1980) are usually one tail. Hence, the 2.5% probability column is used to obtain the 5% probability.

8.4.2 Interpretation

When the unsupplemented and S_1 data are compared (Table 8.1), a value of $F = 1.03$ was obtained. This value is less than the F value in the 2.5% column ($P > 0.05$), and, consequently, there is no evidence that the addition of the medium S_1 increased or decreased the variance in replicate flasks.

8.5 ANALYSIS OF THREE OR MORE GROUPS: BARTLETT'S TEST

8.5.1 How Is the Analysis Carried Out?

If there are three or more groups, then pairs of groups could be tested in turn using the F test described in Section 8.4. A better method, however, is to test all the variances simultaneously using Bartlett's test (Bartlett, 1937; Snedecor & Cochran, 1980). If there are equal numbers of observations in each group, calculation of the test statistic is straightforward and a worked example is shown in Table 8.2. If the three variances do not differ from each other, then the ratio M/C is a member of the χ^2 distribution with $(a - 1)$ DF, where a is the number of groups being compared. If the groups have different numbers of observations in each (unequal n), then the calculations are slightly more complex and are given in Snedecor and Cochran (1980).

TABLE 8.2   Comparison of Variances of Three Groups in Table 8.1 with Equal Observations ($n = 25$) in Each Group Using Bartlett's Test

Group	Variance	ln (variance)
Unsupplemented	463.36	6.1385
S_1	447.88	6.1045
S_2	18,695.24	9.8360
Total	19,606.48	22.079

1. Calculate mean variance = 6,535.49.
2. ln ($\log_e$) mean variance = 8.785.
3. Calculate $M = v\left[a\left(\ln s^{*2}\right) - \sum \ln s_i^2\right]$ where s^{*2} is the mean of the variances, a the number of groups, v = DF of each group. Hence, $M = 102.62$.
4. Calculate $C = 1 + (a + 1)/(3av) = 1.018$.
5. Calculate $\chi^2 = M/C = 102.62/1.018 = 100.8$ (DF $= a - 1$, $P < 0.001$).

8.5.2 Interpretation

In the example in Table 8.2, the value of χ^2 was highly significant ($P < 0.001$), suggesting real differences between the variances of the three groups. The F test carried out in Section 8.4 suggested, however, that the variance of the unsupplemented data was similar to that of the growth medium S_1. Therefore, it is the effect of the growth medium S_2 that has substantially increased the variance of bacterial biomass. Hence, if these data were to be analyzed using ANOVA, the assumption of homogeneity of variance would not hold and a transformation of the data to logarithms to stabilize the variance would be recommended (see Statnote 4).

The use of the χ^2 distribution to test the significance of M/C is questionable if the DF within the groups is less than 5 and in such a case there are special tables for calculating the significance of the statistic (Pearson & Hartley, 1954). Bartlett's test is used less today and may not even be available as part of a software package. This is because the test is regarded as being too "sensitive," resulting in spurious significant results especially with data from long-tailed distributions (Snedecor & Cochran, 1980). Hence, use of the test may raise unjustified concerns about whether the data conform to the assumption of homogeneity of variance. As a consequence, Levene (1960) developed a more robust test to compare three or more variances (Snedecor & Cochran, 1980).

8.6 ANALYSIS OF THREE OR MORE GROUPS: LEVENE'S TEST

8.6.1 How Is the Analysis Carried Out?

Levene's test (Levene, 1960) makes use of the absolute deviations of the individual measurements from their group means rather than their variances to measure the variability within a group. Avoiding the squaring of deviations, as in the calculation of variance, results in a measure of variability that is less sensitive to the presence of a long-tailed distribution. An ANOVA is then performed on the absolute deviations and if significant, the hypothesis of homogeneous variances is rejected.

8.6.2 Interpretation

Levene's test on the data in Table 8.1 using STATISTICA software, for example, gave a value of $F = 52.86$ (DF 2.72; $P < 0.001$) confirming the results of the Bartlett test.

Levene's test has also been called into question since the absolute deviations from the group means are likely to be highly skewed and, therefore, violate another assumption required for an ANOVA, that of normality (see Statnote 6). This problem becomes particularly acute if there are unequal numbers of observations in the various groups being compared. As a consequence, a modification of the Levene test was proposed by Brown and Forsythe (1974).

8.7 ANALYSIS OF THREE OR MORE GROUPS: BROWN–FORSYTHE TEST

The Brown–Forsythe test differs from Levene's test in that an ANOVA is performed not on the absolute deviations from the group *means* but on deviations from the group *medians*. This test may be more accurate than Levene's test even when the data deviate from a normal distribution. Nevertheless, both Levene's and the Brown–Forsythe tests suffer from the same defect in that to assess differences in variance requires an ANOVA, and an ANOVA requires the assumption of homogeneity of variance, which some authors consider to be a fatal flaw of these analyses.

8.8 CONCLUSION

There may be circumstances in which it is necessary for microbiologists to compare variances rather than means, for example, in analyzing data from experiments to determine whether a particular treatment alters the degree of variability or testing the assumption of homogeneity of variance prior to other statistical tests. All of the tests described in this statnote have their limitations. Bartlett's test may be too sensitive, but Levene's and the Brown–Forsythe tests also have problems. We would recommend the use of the variance ratio test to compare two variances and the careful application of Bartlett's test if there are more than two groups. Considering that these tests are not particularly robust, it should be remembered that the homogeneity of variance assumption is usually the least important of those to be considered when carrying out an ANOVA (Snedecor & Cochran, 1980). If there is concern about this assumption and especially if the other assumptions of the analysis are also likely not to be met, for example, lack of normality or nonadditivity of treatment effects (see Statnote 6), then it may be better either to transform the data or to carry out a nonparametric test. Nonparametric analysis of variance will be described in Statnote 24.

Statnote 9

STATISTICAL POWER AND SAMPLE SIZE

Sample size for comparing two independent treatments.
Implications of sample size calculations.
Calculation of power of a test.
Power and sample size in different experimental designs.
Power and sample size in ANOVA.

9.1 INTRODUCTION

There are two important questions that should be asked about any experiment. First, before the experiment is carried out, what sample size (N) would it be appropriate to use to detect a certain "effect"? Second, what is the strength or "power" (P') of an experiment that has been conducted, that is, what difference between two or more groups was the experiment actually capable of detecting? The second question is of particular interest because an experiment in which a nonsignificant difference is reported confirms the H_0 that no difference exists between the groups. This may not mean, however, that the H_0 should actually be rejected because the experiment may have been too small to detect the "true" difference. In any hypothesis test, the statistical method, for example, a t or an F test, indicates the probability of a result if H_0 were actually true, and, therefore, if that probability is less than 5% ($P < 0.05$), H_0 is usually rejected. The ability of an experiment to reject the hypothesis depends on a number of factors, including the probability chosen to reject

Statistical Analysis in Microbiology: Statnotes, Edited by Richard A. Armstrong and Anthony C. Hilton
Copyright © 2010 John Wiley & Sons, Inc.

H_0 (usually set at 0.05), the variability of the measurements, the sample size since larger values of N lead to more accurate estimates of statistical parameters, and the effect size, that is, the size of the actual effect in the population, larger effects being easier to detect. Statistical software (such as GPower) is now widely available to calculate P' and to estimate N in a variety of circumstances, and it is therefore important to understand the value and limitations of this information. This statnote discusses statistical power and sample size as it relates to the comparison of the means of two or more independent groups using a t test or one-way analysis of variance (ANOVA).

9.2 CALCULATE SAMPLE SIZE FOR COMPARING TWO INDEPENDENT TREATMENTS

9.2.1 Scenario

In Statnote 2, an experiment to investigate the efficacy of a novel media supplement in promoting the development of cell biomass was described. Essentially, two sets of 25, 10-liter fermentation vessels were filled with identical growth media with the exception that the media in one of the vessels was supplemented with 10 ml of the novel compound under investigation. The vessels were then inoculated with a culture of a bacterium and the fermentation allowed to proceed until all the available nutrients had been exhausted and growth had ceased. The dry weight of cells was measured in each flask. How many flasks should actually have been used in this experiment?

9.2.2 How Is the Analysis Carried Out?

As a first step, decide on a value δ that represents the size of the difference between the media with and without supplement that is regarded as important and that the experiment is designed to detect. If the true difference is as large as δ, then the experiment should have a high probability of detecting this difference, that is, the test should have a high P' when the true difference is δ. Levels of P' of 0.8 (80%) or 0.9 (90%) are commonly used, whereas levels of 0.95 or 0.99 can be set but are often associated with substantial sample sizes. To determine N for two independent treatments, the following data are required:

1. The size of the difference, δ to be detected
2. The desired probability of obtaining a significant result if the true difference is $\delta (Z_\beta)$ obtained from z tables
3. The significance level of the test (Z_α usually $\rho = 0.05$)
4. The population standard deviation σ usually estimated from previous experiments

The formula for calculating sample size (N) is:

$$N = \frac{(Z_\alpha + Z_\beta)^2 2\sigma^2}{\delta^2} \tag{9.1}$$

Calculation of N using this formula applied to the data presented in Statnote 2 is shown in Table 9.1 and suggests that given the parameters selected, the investigator should have used $N = 36$ in each group to have had an 80% chance of detecting a difference of 10 units. Note that Z_α is based on a two-tail probability but Z_β is always based on a one-sided

TABLE 9.1 Examples of Sample Size (N) and Power (P') Calculation for Comparing Two Independent Treatments

(a) Sample Size Calculation (N)
Difference to be detected $\delta = 10$ units
Standard deviation $\sigma = 15$ units
Significance of test $P = 0.05$, Z_α (from Z table at $P = 0.05$) = 1.96 (two-tail test)
Power of test say $P = 0.80$ and therefore P of **not** demonstrating an effect = 0.20
Z_β = (from Z table at $P = 0.20$) = 0.84 (one-tail test)
$N = \left(Z_\alpha + Z_\beta\right)^2 2\sigma^2 / \delta^2 = 3528/1000 = 35.28$, say 36 per group

(b) Power Calculation (P')
Suppose in example above, the experiment had been carried out with 36 individuals per group but the standard deviation had been 20 units not 15:

$$Z_\beta = \left(\frac{\sqrt{N}\delta}{\sqrt{2}\sigma}\right) - Z_\alpha = 0.17$$

Hence, P of **not** demonstrating an effect = 0.43 (from Z table) and, therefore, experiment has a $P' = 0.57$ (57%) of demonstrating a difference of 10 units.

test (Norman & Streiner, 1994). This is because the tails of the two distributions representing the two media overlap on only one side.

9.3 IMPLICATIONS OF SAMPLE SIZE CALCULATIONS

This procedure is designed to protect the investigator against finding a non-significant result and reporting that the data are consistent with H_0 when in fact the experiment was too small. This suggests that a sample size calculation should always be carried out in the planning stage of an experiment. However, in reality, sample sizes are usually constrained by expense, time, or availability of human subjects for research and quite often a sample size calculation will result in an unrealistic estimate of N. Microbiologists would be surprised at the number of samples required to detect modest differences between two groups given the level of variability often encountered in experiments. Hence, sample size calculations may be an interesting adjunct to a study and may provide an approximate guide to N but should not be taken too seriously (Norman & Streiner, 1994). In addition, increasing sample size is only one method of increasing P'. Reducing the variability between replicate samples by using more homogenous groups or the use of experimental designs such as a paired or randomized block design and which eliminate certain sources of variability may also increase P'. A summary of sample size calculations applied to the most important statistical tests described in this book is presented in Appendix 4.

9.4 CALCULATION OF POWER (P') OF A TEST

9.4.1 How Is the Analysis Carried Out?

Sample size calculations also contain a useful corollary, calculation of the strength, or P' of an experiment to detect a specific difference. This type of calculation is very useful in experiments that have failed to detect a difference the investigator thought was present. In such circumstances, it is useful to ask whether the experiment had sufficient P' to detect the anticipated difference. To calculate P' of an experiment Eq. (9.1) is rearranged to give Z_β:

$$Z_\beta = \left(\frac{\sqrt{N}\delta}{\sqrt{2}\sigma} \right) - Z_\alpha \tag{9.2}$$

9.4.2 Interpretation

A worked example utilizing this equation is given in Table 9.1. Suppose that the experiment described in the previous section had been conducted with a sample size of $N = 36$ but that the σ was actually 20 and not 10 units. The value of Z_β has consequently fallen to 0.17, corresponding to a probability of not demonstrating an effect of $P = 0.43$. Hence, the probability of detecting a difference between the two means of 10 units has fallen to 57% and, hence, P' would have been too low for this experiment to have had much chance of success.

9.5 POWER AND SAMPLE SIZE IN OTHER DESIGNS

The equations used for calculating P' and N differ and depend on the experimental design, for example, in a "paired" design (see Statnote 3) or when comparing two proportions (Katz, 1997). Statistical software is available for calculating P' and N in most circumstances, and, although the equations may differ, the principles described in this statnote remain the same. However, the situation becomes more complicated if there are more than two groups in a study and if the data are analyzed using ANOVA.

9.6 SAMPLE SIZE AND POWER IN ANOVA

Calculation of P' is more complex when several group means are involved because the differences between the means may be distributed in various ways (Fig. 9.1). An important statistic when several means are present is the effect size (d) where $d = \delta/\sigma$ and δ is the difference between the highest and lowest mean (Norman & Streiner, 1994). There are actually various methods of estimating the effect size of increasing levels of complexity, and the present example is one of the simplest. For example, if there are five groups ($K = 5$, a control and four treatments), one treatment may have a large effect while the remaining three may have similar but lesser effects (scenario A). In scenario B, the treatment means are spread more or less evenly, and in scenario C, three treatments have large but similar effects and one has little effect and is therefore similar to the control. The essential approach is that d is transformed into the effect size for ANOVA by multiplying by a formula that varies depending on the distribution of means. Various scenarios and sample formulas are illustrated in Figure 9.1 and how the calculations are made is shown in Table 9.2.

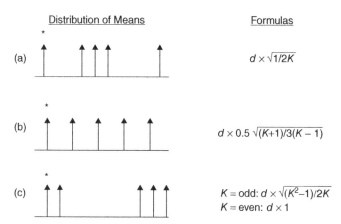

Figure 9.1. Adjustment to the effect size for calculation of sample size in a one-way analysis of variance (ANOVA). In scenario A, one treatment may have a large effect, while the remaining three may have similar but lesser effects. In scenario B, the treatment means are spread more or less evenly, and in scenario C, three treatments have large but similar effects and one has little effect and is therefore similar to the control. (K = number of groups, * = control group).

TABLE 9.2   Sample Size Calculation for One-Way Analysis of Variance (ANOVA)

Difference to be detected (largest mean − smallest mean) = δ
Assume individual means (K groups) equally distributed
Standard deviation = σ

1. Calculate effect size $d = \delta/\sigma$.
2. Adjustment to formula (from Fig. 9.1): effect size for ANOVA $= (f) = \sqrt{d} \times 0.5(K+1)/3(K-1)$.
3. Look up f in Table I (Norman & Streiner, 1994) to give sample size having chosen Z_α and Z_β

9.7   MORE COMPLEX EXPERIMENTAL DESIGNS

In more complex experimental designs (see Statnotes 12 to 14), where there are many treatments or if a factorial arrangement of treatments is present, calculation of N by these methods becomes less useful. A more relevant concept may be to consider the number of DF associated with the error term of the ANOVA. In the general case, in a one-way design (Statnote 6) if there are p treatments and N observations in each group, the error term will have $p(N-1)$ DF, and the greater the value of N, the greater the DF of the error term and the more precise and reliable the error estimate will be. A change of 1 DF has a large effect on t or F when DF < 10, but the effect is quite small when DF > 20. Hence, it is good practice to have at least 15 DF for the error term, and this figure will be dependent on both the number of treatments and N (Ridgeman, 1975). In more complex factorial designs (see Statnotes 12 and 13), with different factors or variables in the experiment, the presence of factorial combinations of treatments leads to internal replication, and, therefore, such experiments can often be carried out by using much smaller numbers of replicates per treatment combination. The principles underlying factorial experiments will be discussed in Statnote 12.

9.8   SIMPLE RULE OF THUMB

Sample size calculations, although they appear to be precise, are only approximations (Norman & Streiner, 1994) and often the simple rule of thumb suggested by Lehr (1992) may provide an adequate estimate of N. For example, Lehr (1992) suggested that for the t test, the equation for N can be simplified as follows:

$$N = \frac{16s^2}{d^2}$$
(9.3)

where d is the difference between the two means and s is the joint SD. Moreover, in one-way or factorial ANOVAs, a possible approach is to choose the difference between two means that is the most important of those to be tested and apply Eq. (9.3).

9.9   CONCLUSION

Statistical software is now commonly available to calculate P' and N for most experimental designs. In many circumstances, however, sample size is constrained by lack of time, cost, and in research involving human subjects, the problems of recruiting suitable individuals. In addition, the calculation of N is often based on erroneous assumptions about σ and, therefore, such estimates are often inaccurate. At best, we would suggest that such calculations provide only a very rough guide of how to proceed in an experiment, and Lehr's rule or the 15 DF rule may be adequate for many purposes. Nevertheless, calculation of P' is very useful especially in experiments that have failed to detect a difference that the experimenter thought was present. We would recommend that P' should always be calculated in such circumstances to determine whether the experiment was actually too small to test the H_0 adequately.

Statnote 10

ONE-WAY ANALYSIS OF VARIANCE (RANDOM EFFECTS MODEL): THE NESTED OR HIERARCHICAL DESIGN

Linear models.

The one-way ANOVA, random-effects model.

How to distinguish random- and fixed-effects factors.

10.1 INTRODUCTION

In Statnote 6, a one-way analysis of variance (one-way ANOVA) in a randomized design was described. In a one-way ANOVA, an individual observation is classified according to which group or treatment it belongs, and observations within each group are a random sample of the relevant population. The scenario to illustrate this analysis compared the degree of bacterial contamination on 2p (pence) coins collected from three types of business premises, namely a butcher shop, a sandwich shop, and a newsstand. A sample of four coins was collected at random from each location and the number of bacterial colonies present on each coin was estimated. The ANOVA appropriate to this design is an example of a *fixed-effects model* because the objective is to estimate the differences between the three businesses, which are regarded as *fixed* or *discrete* effects. There is, however, an alternative model called the *random-effects model* in which the objective is not to measure a fixed effect but to estimate the degree of variation of a particular measurement and to compare different sources of variation in space and/or time. These designs are often called *nested* or *hierarchical* designs (Snedecor & Cochran, 1980).

Statistical Analysis in Microbiology: Statnotes, Edited by Richard A. Armstrong and Anthony C. Hilton
Copyright © 2010 John Wiley & Sons, Inc.

10.2   SCENARIO

The contribution of hands contaminated with pathogenic microorganisms to the spread of infectious disease has been recognized for many years. Of particular importance are communal areas where shared facilities of a tactile nature may present an increased opportunity for cross contamination of the fingers. A study was, therefore, undertaken to determine the role of computer keyboards in a university communal computer laboratory as a source of microbial contamination of the hands. The data presented in this statnote relate to a component of the study to determine the aerobic colony count (ACC) of 10 selected keyboards with samples taken from two keys per keyboard determined at two times a day, namely at 9 a.m. and 5 p.m. Ten keyboards were selected randomly from those available in the computer laboratory and samples taken from two keys per keyboard (the a and z keys) using a cotton swab moistened in sterile distilled water (SDW). The swab was returned to 1 ml of SDW and the swab agitated to release the microorganisms recovered from the surface into the liquid. A 0.1-ml sample of the SDW was plated onto nutrient agar and incubated at 30°C for 24 hours following which the colony forming units (CFUs) per milliliter was calculated.

10.3   DATA

The data form a hierarchical classification with three levels, namely keyboards, keys within a keyboard, and time of day. The data obtained are presented in Table 10.1.

10.4   ANALYSIS

10.4.1   How Is the Analysis Carried Out?

There is a commonly used notation to describe the basic model of an ANOVA (Snedecor & Cochran, 1980; see Statnote 6). The subscript i is used to denote the group or class (i.e., the treatment group), i taking the values 1 to a, whereas the subscript j designates the members of the class, j taking the values 1 to n (hence, a groups and n replicates or observations per group). Within class i, the observations x_{ij} are assumed to be normally distributed about a mean μ with variance σ^2. This is an example of a *linear model* and can be written thus:

$$x_{ij} = \mu + a_i + e_{ij} \qquad (10.1)$$

Hence, any observed value x_{ij} is the sum of three parts: (1) the overall mean of all the observations μ, (2) a treatment or class deviation a_i, and (3) a random element e_{ij} taken from a normally distributed population. The random element reflects the combined effects of natural variation that exists between replications and errors of measurement. The more complex types of ANOVA can be derived from this simple linear model by the addition of one or more further terms.

10.4.2   Random-Effects Model

Equation (10.1) describes essentially a fixed-effects model in which the a_i are fixed quantities to be estimated. The corresponding random-effects model is similar, but the

TABLE 10.1 Aerobic Colony Count Recovered from the *a* and *z* Keys of 10 Computer Keyboards in Communal Use Sampled at 9 a.m. and 5 p.m.

Keyboard	Key	Time	CFU/ml
Keyboard 1	*a*	a.m.	170
		p.m.	210
	z	a.m.	55
		p.m.	90
Keyboard 2	*a*	a.m.	437
		p.m.	450
	z	a.m.	200
		p.m.	179
Keyboard 3	*a*	a.m.	210
		p.m.	350
	z	a.m.	5
		p.m.	140
Keyboard 4	*a*	a.m.	560
		p.m.	470
	z	a.m.	10
		p.m.	93
Keyboard 5	*a*	a.m.	47
		p.m.	166
	z	a.m.	12
		p.m.	63
Keyboard 6	*a*	a.m.	921
		p.m.	1043
	z	a.m.	237
		p.m.	178
Keyboard 7	*a*	a.m.	34
		p.m.	21
	z	a.m.	0
		p.m.	8
Keyboard 8	*a*	a.m.	585
		p.m.	658
	z	a.m.	34
		p.m.	67
Keyboard 9	*a*	a.m.	647
		p.m.	457
	z	a.m.	34
		p.m.	56
Keyboard 10	*a*	a.m.	78
		p.m.	67
	z	a.m.	24
		p.m.	3

symbols KB, (representing keyboards) and K (representing keys) are included. The difference between this model and Eq. (10.1) is that KB_i and K_{ij} are considered to be random variables, and the term e_{ijk} refers to errors of measurement and to the fact that microbial content is determined on two occasions (a.m. and p.m.). This model can be written thus:

$$x_{ij} = \mu + \text{KB}_i + K_{ij} + e_{ijk} \qquad\qquad (10.2)$$

10.4.3   Interpretation

The ANOVA is shown in Table 10.2. In a random-effects model, it is possible to calculate the *components of variance* (sample estimates of which are given the symbol s^2, and population values σ^2), and these are often more informative than the usual F tests. The components of variance are estimates of the variance of the measurements made between keyboards (σ_{KB}^2), between keys within a keyboard (σ_K^2), and between determinations (a.m./p.m.) within an individual key (σ_D^2) and can be calculated from the ANOVA. In the example quoted, the analysis suggested that the variance between keyboards was essentially zero compared with that attributable to keys (74707.6), which, in turn, was more than 20 times that attributed to the variation between determinations made on the same key in the morning and afternoon (3379.33).

This experiment provides two important pieces of information. First, there is little significant variation between the different keyboards or between measurements made in the morning and the afternoon compared with that between individual keys. This result suggests that in a future study, designed to estimate the degree of microbial contamination on keyboards, a simpler sampling strategy could be employed involving fewer keyboards and a single sample time. Second, the difference in microbial contamination of the two keys is substantial, and, therefore, to improve the accuracy of estimates of contamination of a keyboard as a whole, more keys should be sampled. Although the experiment was not designed to test the difference between specific keys, the results suggest the hypothesis that a more frequently used key such as a may have a considerably greater degree of contamination than the more rarely used z key, and this hypothesis could be tested by a more rigorous experiment. These results emphasize the usefulness of the random-effects model in preliminary experiments designed to estimate different sources of variation and to plan an appropriate design.

TABLE 10.2   A One-Way Analysis of Variance (ANOVA) (Nested Experimental Design with Three Levels) of Data in Table 10.1[a]

Variation	DF	SS	MS (s^2) σ^2Estimated
1. Keyboards	9	1,110,632.23	$123{,}403.581 = \sigma_D^2 + 2\sigma_K^2 + 4\sigma_{\text{KB}}^2$
2. Keys within keyboards	10	1,527,945.25	$152{,}794.525 = \sigma_D^2 + 2\sigma_K^2$
3. Determinations	20	67,586.5	$3{,}379.33 = \sigma_D^2$

Components of Variance	Estimated Variance
Between keyboards (σ_{KB}^2)	0
Between keys within a keyboard (σ_K^2)	74,707.6
Between am/pm within a key (σ_D^2)	3,379.33

[a] DF = degrees of freedom; SS = sums of squares; MS = mean square.

10.5 DISTINGUISH RANDOM- AND FIXED-EFFECT FACTORS

It is necessary to be able to identify whether a fixed- or random-effect model is the most appropriate in each experimental context. This is essential in more complex factorial-type designs in which there may be a mixture of both fixed- and random-effect factors (mixed models) (Snedecor & Cochran, 1980). One way of deciding whether a factor is fixed or random is to imagine the effect of changing one of the levels of the factor (Ridgman, 1975). If this substantially alters the experiment, for example, by substituting a confectioners shop for the newsstand in Statnote 6, then it is a fixed-effect factor. By contrast, if we considered it the same experiment, for example, substituting a different keyboard or key would have little effect on the overall objectives of the experiment, then it would be a random-effect factor. Hence, a random-effect factor is only a sample of the possible levels of the factor and the intent is to generalize to all levels, whereas a fixed factor contains all levels of the factor that are of interest in the experimental design (Norman & Streiner, 1994). Whether a particular factor is considered to be random or fixed may also depend on context. For example, the two keys measured in the present scenario were originally regarded as a sample of the population of keys on the keyboard. However, having selected the a and z key and found a significant component of variance associated with them, one could envisage an experiment to investigate the specific difference between such keys. In this new experiment, we would deliberately want to study the a and z keys and the factor *key* would now become a fixed-effect factor.

10.6 CONCLUSION

There is an alternative model of the one-way ANOVA called the random-effects model or nested design in which the objective is not to determine the significance of an effect but to estimate the degree of variation of a particular measurement and to compare different sources of variation that influence the measurement in space and/or time. The most important statistics from a random-effects model are the components of variance that estimate the variance associated with each of the sources of variation influencing a measurement. The nested design is particularly useful in preliminary experiments designed to estimate different sources of variation and in the design of experiments.

Statnote 11

TWO-WAY ANALYSIS OF VARIANCE

The two-way design.
Statistical model of the two-way design.
Advantages of a two-way design.
Comparison of one-way and two-way designs.

11.1 INTRODUCTION

In Statnote 6, a one-way analysis of variance (one-way ANOVA) was described in which an individual observation was classified according to which group or treatment it belonged. The scenario to illustrate this analysis compared the degree of bacterial contamination on 2p (pence) coins collected from three types of business premises, namely a butcher shop, a sandwich shop, and a newsstand. A sample of four coins was collected at random from each location, and the number of bacterial colonies present on each coin was estimated. Such an experiment is often described as in a *randomized design* as replicates are either assigned at random to treatments or observations within each group are an independent random sample of the relevant population. More complex experimental designs are possible, however, in which an observation may be classified in two or more ways (Snedecor & Cochran, 1980).

Statistical Analysis in Microbiology: Statnotes, Edited by Richard A. Armstrong and Anthony C. Hilton
Copyright © 2010 John Wiley & Sons, Inc.

11.2 SCENARIO

We return to the scenario described in Statnote 8. An experiment was carried out to investigate the efficacy of two novel media supplements (S_1 and S_2) in promoting the development of cell biomass. On each of 10 occasions, three 10-liter fermentation vessels were sterilized and filled with identical growth media with the exception that the media in two of the vessels was supplemented with 10 ml of either medium supplement S_1 or S_2. The vessels were allowed to equilibrate and were subject to identical environmental/incubation conditions. The vessels were then inoculated with a culture of a bacterium at an equal culture density, and the fermentation was allowed to proceed until all the available nutrients had been exhausted and bacterial growth had ceased. The entire volume of culture media in each fermentation vessel was then removed and filtered to recover the bacterial biomass, which was subsequently dried and the dry weight of cells measured.

11.3 DATA

This experiment could have been carried out in a completely randomized design, that is, the replicate vessels were allocated at random to the treatments and analyzed using a one-way ANOVA (see Statnote 6). The present experiment, however, was carried out using a different method. First, 30 vessels were divided into 10 groups of 3-, each group of 3 representing a *replication* with the intention of setting up and processing each replication (a control and each of the two treatments S_1, S_2) on each of 10 separate occasions. Second, the treatments were allocated to the 3 vessels within a replication independently and at random. The analysis appropriate to this design is an extension to that of the paired sample *t* test described in Statnote 3 but applied to more than 2 treatments. The data are presented in Table 11.1.

11.4 ANALYSIS

11.4.1 How Is the Analysis Carried Out?

In a two-way design, each treatment is allocated by randomization to one experimental unit within each group. The name given to each group varies with the type of experiment.

TABLE 11.1 Effect of Two Novel Media Supplements (S_1, S_2) on Bacterial Biomass Measured on 10 Separate Occasions

Occasion (Blocks)	Control	$+S_1$	$+S_2$
1	461	562	354
2	472	573	359
3	473	574	369
4	481	581	403
5	482	582	425
6	494	586	476
7	493	591	511
8	495	592	513
9	506	592	556
10	502	607	578

Originally, the term *randomized blocks* was applied to this type of design because it was first used in agricultural experiments in which experimental treatments were given to units within "blocks" of land, plots within a block tending to respond more similarly compared with plots in different blocks (Snedecor & Cochran, 1980). In the present example, the block is a single trial, or replication, of the comparison between treatments, the trial being carried out on 10 separate occasions. In a further application, in experiments with human subjects, there is often considerable variation from one individual to another, and hence a good strategy is to give all treatments successively to each "subject" in a random order, the subject therefore comprising the block or replication.

11.4.2 Statistical Model of Two-Way Design

In the example given in Table 11.1, the fact that the vessels are also grouped into replications, 1 complete replication for each of the 10 occasions, gives a more complex model. Using the commonly used notation to describe the basic model of an ANOVA first described in Statnote 10, the two-way design includes a term for the replication effect b in addition to the treatment effect a, namely

$$x_{ij} = \mu + a_i + b_j + e_{ij} \tag{11.1}$$

Hence, the ANOVA table (Table 11.2) includes an extra term for replications, that is, the variation between occasions. In the terminology used in Statnote 10, treatment a_i is a fixed-effect factor whereas blocks or occasions b_j are a random-effect factor. In addition to the assumptions made in the randomized design, namely homogeneity of variance, additive class effects, and normal distribution of errors, this type of design makes the additional assumption that the difference between treatments is consistent across all replications (Snedecor & Cochran, 1980).

11.4.3 Interpretation

The ANOVA appropriate to the two-way design is shown in Table 11.2. This design is often used to remove the effect of a particular source of variation from the analysis. For example, if there was significant variation due to replications and if treatments had been allocated to vessels at random, then all of the *between replicates* variation would have been included in the pooled error variance. The effect of this would be to increase the

TABLE 11.2   Analysis of Variance (Two-Way in Randomized Blocks) of Data in Table 11.1[a]

		ANOVA table		
Variation	DF	SS	MS	F
Treatments	2	91,373.4	45,686.7	27.32[b]
Replications	9	35,896.0	3,988.45	2.39[c]
Error	18	30,097.3	1,672.07	

[a] DF = degrees of freedom; SS = sums of squares; MS = mean square, F = variance ratio.
[b] $P < 0.001$.
[c] Not quite significant at $P = 0.05$; post hoc tests (Scheffé): control compared with S_1, $S = 14.39$ ($P < 0.05$); control compared with S_2, $S = 1.48$ ($P > 0.05$), S_1 compared with S_2, $S = 25.11$ ($P < 0.05$).

error variance and to reduce the P' of the experiment (see Statnote 9), thus making it more difficult to demonstrate a possible treatment effect. In a two-way design, however, variation between replications attributable to occasions is calculated as a separate effect and, therefore, does not appear in the error variance. This may increase the P' of the experiment and make it more probable that a treatment effect would be demonstrated. In the example quoted (Table 11.2), there is a highly significant effect of media supplement ($F = 27.32$, $P < 0.001$). In a two-way design, planned comparisons between the means or post hoc tests can be performed as for the randomized design (see Statnote 7). Hence, Scheffé's post hoc test (see Statnote 7) suggested that this result is largely due to the effect of supplement S_1 increasing yield. In addition, in a two-way design, the variation due to replications is calculated ($F = 2.39$), and this was not quite significant at $P = 0.05$. The borderline significance suggests there may have been some differences between replications, and removing this source of variation may have increased the P' of the experiment to some degree. This statnote also illustrates an alternative method of increasing the P' of an experiment, namely by altering its design.

A comparison of the ANOVA table in Table 11.1 with that for a one-way ANOVA in a randomized design demonstrates that reducing the error variance by "blocking" has a cost, namely a reduction in the DF of the error variance, which makes the estimate of the error variation less reliable. Hence, an experiment in a two-way design would only be effective if the blocking by occasion or some other factor reduced the pooled error variance sufficiently to counter the reduction in DF (Cochran & Cox, 1957; Snedecor & Cochran, 1980).

11.5 CONCLUSION

The two-way design has been variously described as a matched-sample F test, a simple within-subjects ANOVA, a one-way within-groups ANOVA, a simple correlated-groups ANOVA, and a one-factor repeated measures design! This confusion of terms is likely to lead to problems in correctly identifying this analysis within commercially available packages. The essential feature of the design is that each treatment is allocated by randomization to one experimental unit within each group or block. The block may be a plot of land, a single occasion in which the experiment was performed, or a human subject. The blocking is designed to remove an aspect of the error variation and increase the P' of the experiment. If there is no significant source of variation associated with the blocking, then there is a disadvantage to the two-way design because there is a reduction in the DF of the error term compared with a fully randomized design, thus reducing the P' of the analysis.

Statnote 12

TWO-FACTOR ANALYSIS OF VARIANCE

The factorial experimental design.

Interactions between variables.

Statistical model of a factorial design.

Partitioning the treatments sums of squares into factorial effects (contrasts).

12.1 INTRODUCTION

The analyses of variance (ANOVA) described in previous statnotes (see Statnotes 6, 10, and 11) are examples of single-factor experiments. The *single-factor experiment* comprises two or more treatments or groups often arranged in a randomized design, that is, experimental units or replicates are assigned at random and without restriction to the treatments. An extension to this experimental design is the two-way design (see Statnote 11) in which the data are classified according to two criteria, that is, *treatment* or *group* and the *replicate* or *block* to which the treatment belongs. A factorial experiment, however, differs from a single-factor experiment in that the effects of two or more factors or "variables" can be studied at the same time. Combining factors in a single experiment has several advantages. First, a factorial experiment usually requires fewer replications than an experiment that studies each factor individually in a separate experiment. Second, variation between treatment combinations can be broken down into components representing specific comparisons or *contrasts* (Ridgman, 1975; Armstrong & Hilton, 2004) that reveal the possible

Statistical Analysis in Microbiology: Statnotes, Edited by Richard A. Armstrong and Anthony C. Hilton
Copyright © 2010 John Wiley & Sons, Inc.

synergistic or interactive effects between the factors. The interactions between factors often provide the most interesting information from a factorial experiment and cannot be obtained from a single-factor experiment. Third, in a factorial design, an experimenter can often add variables considered to have an uncertain or peripheral importance to the design with little extra effort. This statnote describes the simplest case of a factorial experiment incorporating two factors each present at two levels.

12.2 SCENARIO

The kitchen dishcloth is increasingly recognized as a primary reservoir of bacteria with potential to cause widespread cross contamination in food preparation environments. An investigator wished to study the influence of the material from which the dishcloth was manufactured (cloth or sponge) (factor A) and the effect of rinsing the dishcloth in running water (factor B) on the number of bacteria subsequently transferred to a food preparation surface (Hilton & Austin, 2000). Dishcloths of each material type were inoculated with 1 ml of a 10^8 CFU/ml *Escherichia coli* culture and after 10 minutes, the cloth was wiped over an appropriate area of sterile cutting board. Additional pieces of cloth were inoculated but rinsed in sterile running water before wiping. The cutting board was subsequently swabbed to recover *E. coli* that had been deposited from the wiping by the cloth.

12.3 DATA

The data comprise the number of bacterial colonies obtained on nutrient agar from the two types of dishcloth, rinsed and unrinsed, and are presented in Table 12.1. The objectives of the experiment were to determine whether the ability of bacteria to be transferred from the dishcloth varied, first, with the type of dishcloth (factor A) and, second, with rinsing treatment (factor B) and whether the two factors had an independent influence on the numbers of bacteria. Hence, there are four treatment combinations, that is, two types of cloth each of which was either rinsed or not rinsed. This type of design is the simplest type of factorial experiment and is also known as a 2^2 factorial, that is, two factors with two levels of each factor. In this notation, the superscript refers to the number of factors or variables included and the integer the number of levels of each factor. As the number of variables included in the experiment increases, the superscript can take any integer value. Hence, a 2^4 factorial would have four separate factors each at two levels.

TABLE 12.1 Influence of Type of Dishcloth and Rinsing Treatment on Number of Bacteria Transferred to Food Preparation Surface

Factor A	Cloth		Sponge	
Factor B	Rinsed	Not rinsed	Rinsed	Not rinsed
	1.0×10^5	7.8×10^7	3.9×10^5	8.0×10^6
	2.3×10^4	5.0×10^7	9.0×10^3	4.0×10^6
	3.9×10^5	4.1×10^7	8.5×10^4	2.0×10^6
Mean	1.7×10^5	5.6×10^7	1.6×10^5	4.7×10^6

12.4 ANALYSIS

12.4.1 How is the Analysis Carried Out?

Using the notation described previously (see Statnote 10), this design can be described as follows:

$$x_{ijk} = \mu + a_i + b_j + (ab)_{ij} + e_{ijk} \tag{12.1}$$

In this model, x_{ijk} is the value of the kth replicate of the ith level of factor A and the jth level of factor B, a_i and b_j are the main effects of each of the two factors, and $(ab)_{ij}$ represents the two-factor interaction between A and B.

As in previous examples, the total sums of squares (SS) of the data can be partitioned into components associated with differences between the effects of the cloth and rinsing treatment (see Statnote 7). In this case, the between-treatments SS can be partitioned into contrasts that describe the "main" effects of A and B and the "interaction" effect $A \times B$. These effects are linear combinations of the treatment means, each being multiplied by a number or coefficient to calculate a particular effect. In fact, the meaning of a factorial effect can often be appreciated by studying these coefficients (Table 12.2). The effect of dishcloth is calculated from those replicates representing the cloth data (+) compared with those that represent the sponge data (−). Note that in a factorial design, every observation is used in the estimate of the effect of every factor. Hence, factorial designs have *internal replication* and this may be an important consideration in deciding the number of replications to use. Each replicate is actually providing two estimates of the difference between cloths and sponges. The main effect of rinsing is calculated similarly to that of dishcloth. By contrast, the two-factor interaction (dishcloth × rinsing) can be interpreted as a test of whether cloth type and rinsing treatment act independently of each other. A significant interaction term would imply that the effect of the combination of cloth type and rinsing treatment would not be predictable from knowing their individual effects; a common occurrence in many real situations in which it is a combination of factors that determine a particular outcome.

In a 2^2 factorial, partitioning the treatments SS into factorial effects provides all the information necessary for interpreting the results of the experiment and further post hoc tests would not be necessary. With more complex factorial designs, for example, those with more than two levels of each factor, further tests may be required to interpret a main effect or an interaction. With factorial designs, however, it is better to define specific comparisons before the experiment is carried out rather than to rely on post hoc tests that

TABLE 12.2   Coefficients for Calculating Factorial Effects or Contrast in a 2^2 Factorial for the Data in Table 12.1

Type of Dishcloth (A)	Cloth		Sponge	
Rinsing (B)	Yes	No	Yes	No
Treatment Total	5.13×10^5	16.9×10^7	4.84×10^5	14.0×10^6
Contrast				
A	+1	+1	−1	−1
B	+1	−1	+1	−1
AB	+1	−1	−1	+1

TABLE 12.3   Analysis of Variance of Dishcloth Data in Table 12.1[a]

Source	DF	SS	MS	F	P
Cloth/sponge	1	2002.83	2002.83	20.99	<0.01
Rinsing	1	2760.42	2760.42	28.92	<0.001
Interaction	1	2001.33	2001.33	20.97	<0.01
Error	8	763.49	95.44		

[a] DF = degrees of freedom, SS = sums of squares, MS = mean square, F = variance ratio, P = Probability.

compare all possible combinations of the means. Factorial experiments can be carried out in a completely randomized design (see Statnote 6), in randomized blocks (see Statnote 11), or in more complex designs (Cochran & Cox, 1957). The relative advantages of these designs are the same as for the one-way design.

12.4.2   Interpretation

The resulting ANOVA (Table 12.3) is more complex than in a one-way experiment because the between-groups or treatments SS is partitioned into three factorial effects, namely the main effects of dishcloth type and rinsing and the interaction between the two factors. In the present example, there is a main effect of dishcloth type ($F = 20.99$, $P < 0.01$) and of rinsing ($F = 28.92$, $P < 0.001$), suggesting significantly more bacteria were transferred from the cloth than the sponge and significantly fewer after rinsing both materials. In addition, there is significant interaction between the factors ($F = 20.97$, $P < 0.01$), suggesting that the effect of rinsing is not consistent for the two types of dishcloth, namely rinsing has less effect on the number of bacteria transmitted from the sponge compared with the cloth. Examination of the treatment means suggests that the sponge transfers a smaller proportion of its bacterial load to the food preparation surface compared with the cloth. This effect is probably attributable to organisms being more exposed on the surface of the cloth and therefore more liable to be transferred compared with the more cavernous sponge (Hilton & Austin, 2000).

12.5   CONCLUSION

Experiments combining different groups or factors are a powerful method of investigation in microbiology. ANOVA enables not only the effects of individual factors to be estimated but also their interactions, information that cannot be obtained readily when factors are investigated separately. In addition, combining different treatments or factors in a single experiment is more efficient and often reduces the number of replications required to estimate treatment effects adequately. Because of the treatment combinations used in a factorial experiment, the DF of the error term in the ANOVA is a more important indicator of the P' of the experiment than simply the number of replicates per treatment (see Statnote 9).

Statnote 13

SPLIT-PLOT ANALYSIS OF VARIANCE

Major and minor factors.
Statistical model of the split-plot design.
How to identify a split-plot design.
Disadvantages of a split-plot design.

13.1   INTRODUCTION

In Statnote 12, an investigator wished to study the influence of type of dishcloth (cloth or sponge) (factor A) and rinsing treatment (factor B) on the number of bacteria transferred to a food preparation surface (Hilton & Austin, 2000). The objectives of this experiment were to determine whether the risk of transferring bacteria from the dishcloth varied with dishcloth type, rinsing treatment, and whether the two factors had an independent influence on the number of bacteria transferred. Hence, there were four treatment combinations, namely, two types of cloth, each of which was either rinsed or not rinsed. This type of design is an example of the simplest factorial experiment, also known as a 2^2 factorial, that is, two factors were present with two levels of each factor. An important feature of this experimental design is that the replicates were assigned at random to all possible combinations of the two factors "without restriction." In some experimental situations, however, the two factors are not equivalent to each other, and replicates cannot be assigned at random to all treatment combinations. A common case, called a *split-plot design*, arises when one factor can be considered to be a major factor and the other a minor factor.

Statistical Analysis in Microbiology: Statnotes, Edited by Richard A. Armstrong and Anthony C. Hilton
Copyright © 2010 John Wiley & Sons, Inc.

13.2 SCENARIO

A microbiologist was interested in the number of zoospores produced by the pathogenic aquatic fungus *Saprolegnia diclina* (Smith et al., 1984). The number of zoospores produced and their motility are markedly affected by such factors of the aquatic environment as pH, oxygen tension, and the presence of biocides. To examine the effect of pH on zoospore production and motility, parent colonies of *S. diclina* were placed at the end of experimental counting channels each consisting of five sequential chambers (A to E) (Fig. 13.1). The channels were filled with sterile 5 μm phosphate buffer and then modified to provide environments of either pH = 5.0 or 7.0. Zoospore activity was determined by counting the number of encysted zoospores within each of the five chambers, representing different distances from the parent colony. Each pH channel was replicated three times.

13.3 DATA

The data are presented in Table 13.1. This experiment also has two factors similar to Statnote 12, namely, variation in pH (5.0 or 7.0) and the distance the zoospores travel along the channel from the parent colony (five positions being sampled from A to E). The problem that arises in this type of design is the dependence or correlation between the measurements made in the different chambers within the same channel. Hence, in this experiment, pH is regarded as the *major* factor being applied to the channel as a whole, whereas distance along the channel is the *minor* factor, representing the chambers or subdivisions of the channel. The obvious difference between this and an ordinary factorial design is that, previously, all treatment combinations were assigned at random to replicates, whereas in a split-plot design replicates can only be assigned at random to the main-plot factor, namely, the channels and not to channel–chamber combinations. In some split-plot designs, experimenters allocate replicates to major factors at random and then assign the levels of the minor factor at random within each major block. In yet other variations of this design, the subplots may be divided further to give a split–split-plot design (Snedecor & Cochran, 1980).

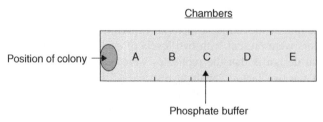

Figure 13.1. Counting chamber used to study the influence of pH on the production and motility of zoospores (number of encysted zoospores mm²/24 hours) of the aquatic fungus *Saprolegnia diclina*.

TABLE 13.1   Influence of pH on Production and Motility of Zoospores (Number of Encysted Zoospores mm²/24 hours) of Aquatic Fungus *Saprolegnia diclina*

Major Factor (Channels)	Minor Factor	Replicate Channels		
		1	2	3
pH = 5.0	Chamber A	3	4	5
	Chamber B	1	2	2
	Chamber C	0	0	1
	Chamber D	1	0	1
	Chamber E	0	0	0
pH = 7.0	Chamber A	20	18	23
	Chamber B	5	7	8
	Chamber C	3	4	5
	Chamber D	2	1	3
	Chamber E	2	1	2

TABLE 13.2   Analysis of Variance of Data in Table 13.1[a]

Source of Variation	Term	DF	SS	MS	F	P
Main plots (pH)	M_i	1	235.200	235.20	67.8	<0.01
Main plot error	e_{ij}	4	13.8667	3.4667		
Subplots (chambers)	T_k	4	522.80	130.70	172.35	<0.001
Interaction	$(MT)_{ik}$	4	229.4667	57.3667	75.65	<0.001
Subplot error	d_{ijk}	16	12.1333	0.7583		

[a] DF = degrees of freedom, SS = sums of squares, MS = mean square, F = variance ratio; P = probability.

13.4   ANALYSIS

13.4.1   How Is the Analysis Carried Out?

The model for a two-factor, split-plot design is

$$x_{ijk} = \mu + M_i + T_k + e_{ij} + (MT)_{ik} + d_{ijk} \tag{13.1}$$

In this case, M represents main-plot treatments and T subplot treatments. The symbols i and j indicate the main plots while k identifies the sub-plot within a main plot. The two components of error e_{ij} and d_{ijk} represent the fact that the error variation between main plots (channels) will be different from that between subplots (chambers). For example, one might expect there to be less natural variation between chambers within a channel than between different channels.

The resulting ANOVA (Table 13.2) is more complex than that for a simple factorial design (Statnote 12) because of the different error terms. Hence, in a two-factor, split-plot ANOVA, two errors are calculated, the main-plot error is used to test the main effect of pH while the subplot error is used to test the main effect of chambers and the possible interaction between the two factors.

13.4.2 Interpretation

There is a significant increase in zoospore production at pH 7.0 compared with pH 5.0 and a marked decline in numbers of zoospores with distance from the parent colony. In addition, there is a significant interaction term, suggesting that the decline in zoospores with distance down the channel is more marked at pH = 7.0 compared with pH = 5.0 (Fig. 13.2). The subplot error is usually smaller that the main-plot error and also has more DF. Hence, such an experimental design will usually estimate the main effect of the subplot factor and its interaction with the main factors more accurately than the main effect of the major factor. Some experimenters will deliberately design an experiment as a split-plot to take advantage of this property.

A disadvantage of a split-plot design is that, occasionally, the main effect of the major factor may be large but not significant, while the main effect of the minor factor and its interaction may be significant but too small to be biologically important. In addition, a common mistake is for researchers to analyze a split-plot design as if it were a fully randomized two-factor experiment. In this case, the single error variance would either be too small or too large for testing the individual treatment effects, and the wrong conclusions could be drawn from the experiment. To decide whether a particular experiment is in a split-plot design, it is useful to consider the following questions: (1) Are the factors equivalent or does one appear to be subordinate to the other and especially does one represent "subdivisions" of the other as in the chambers within a channel? (2) Is there any restriction in how replicates are assigned to the treatment combinations? (3) Is the error variation likely to be the same for each factor?

Caution should also be employed in the use of post hoc tests in the case of a split-plot design. Post hoc tests assume that the observations taken on a given channel are uncorrelated so that the subplot factor group means are not related. This is unlikely since some correlation between measurements made in the different chambers of a channel is inevitable. Standard errors appropriate to the split-plot design can be calculated (Cochran & Cox, 1957; Freese, 1984) but should be used with caution to make specific comparisons

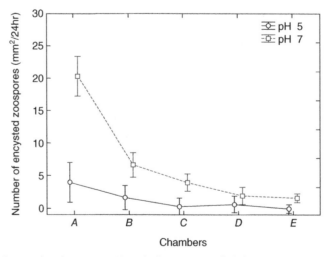

Figure 13.2. Interaction between pH and distance traveled by zoospores down a channel ($F = 75.65$; $P < 0.001$). Error bars are 95% confidence intervals.

between the treatment means. A better method is to partition the SS associated with main effects and interaction into specific contrasts and to test each against the appropriate error (Snedecor & Cochran, 1980).

13.5   CONCLUSION

In some experimental situations, the factors may not be equivalent to each other and replicates cannot be assigned at random to all treatment combinations. A common case, called a split-plot design, arises when one factor can be considered to be a major factor and the other a minor factor. Investigators need to be able to distinguish a split-plot design from a fully randomized design as it is a common mistake for researchers to analyze a split-plot design as if it were a fully randomized factorial experiment.

Statnote 14

REPEATED-MEASURES ANALYSIS OF VARIANCE

A more complex three-factor ANOVA.
The repeated-measures design.
Repeated-measures designs and post hoc tests.

14.1 INTRODUCTION

In Statnote 13, the split-plot experimental design was described as a variation of the fully randomized, factorial ANOVA. An important feature of a fully randomized design is that the experimental units or *replicates* are assigned at random to all possible combinations of the factors. With two factors arranged in a split-plot design, however, replicates cannot be assigned at random to all possible treatment combinations. In a split-plot design, one factor is usually designated as the *major* factor and the other the *minor* factor. Replicates are assigned at random to different levels of the major factor, and then the different levels of the subplot factor may be assigned at random within each of the main plots. In some circumstances, the minor factor is a *subdivision* of the major factor; as in Statnote 13 where the *chambers* (minor factor) can be regarded as subdivisions of a *channel* (major factor). The ANOVA of a split-plot design is also different from that of a completely randomized factorial design in that it incorporates two error terms. One error (the main-plot error) is used to test the significance of the main effect of the major factor and the second (the subplot error) tests the main effect of the minor factor and its interaction with the major factor. A further variation of the split-plot design arises when the subplot factor represents

Statistical Analysis in Microbiology: Statnotes, Edited by Richard A. Armstrong and Anthony C. Hilton
Copyright © 2010 John Wiley & Sons, Inc.

measurements taken on the same replicate or subject sequentially in time and is often referred to as a *repeated-measures* design. This statnote discusses the analysis of a more complex factorial experiment that incorporates three different factors, one of which is a repeated measure.

14.2   SCENARIO

An investigator wished to examine the pattern of survival of bacteria on the surface of £5 notes. Two species, namely, *Escherichia coli* and *Staphylococcus epidermidis* were inoculated onto the surface of a sample of £5 notes, and the numbers of surviving bacteria estimated at 10 time intervals over a subsequent 55-hour period. The experiment was replicated twice. Survival of the two bacteria on glass coverslips over the same period of time was examined as a control. The objectives of the experiment were, first, to determine whether there was a difference in the pattern of survival of bacteria on control surfaces as against £5 notes and, second, to determine whether the two bacterial strains exhibited different patterns of survival over time. Hence, the interactions between the type of surface, bacterial strain, and time are of particular interest in this experiment.

14.3   DATA

The data are presented in Table 14.1. This experiment is more complex than those described in previous statnotes. First, three factors are involved, namely, bacterial species (2 levels, *E. coli* and *S. epidermidis*), type of surface (2 levels, glass coverslips and £5 notes), and time (10 sample times), and this results in a *three-factor* ($2 \times 2 \times 10$) factorial. Second, the fact that repeated measurements of bacterial numbers are made on each surface leads to a repeated-measures design, that is, the same measurement (bacterial number) is made sequentially at specific time intervals on each replicate.

14.4   ANALYSIS

14.4.1   How is the Analysis Carried Out?

There are two major factors (bacteria and type of surface) while time constitutes the repeated-measures factor. As in the split-plot design (see Statnote 13), there are two errors; the main-plot error, which tests the main effects of bacterial strain and type of surface and their interaction, and the subplot error, which tests the main effect of time and its interaction with the other two variables.

14.4.2   Interpretation

An ANOVA of the data combines features of both Statnotes 12 and 13 and is shown in Table 14.2. Hence, there is a significant main effect of bacterial strain ($F = 140.81$, $P < 0.001$), a not surprising finding as there were considerably more *S. epidermidis* overall than *E. coli* on the notes and coverslips. In addition, there is a significant main effect of type of surface ($F = 111.48$, $P < 0.001$), which suggests greater numbers of surviving bacteria, regardless of species, on the £5 notes compared with glass coverslips. More

TABLE 14.1 Survival of *Escherichia coli (EC)* and *Staphylococcus epidermidis* (SE) on £5 Notes and Glass Coverslips (Control) with Two Replications[a]

	Notes		Control		Notes		Control	
Time (hours)	EC	EC	EC	EC	SE	SE	SE	SE
0	3800	3800	3800	3800	5400	5400	5400	5400
1	800	430	0	0	600	800	7	22
3	500	351	6	0	560	560	6	10
5.5	446	249	1	0	700	764	0	2
24	0	1	0	0	272	171	0	0
27	0	0	0	0	54	2	0	0
30	0	0	0	0	79	42	0	0
48	0	0	0	0	124	105	0	0
51	0	0	0	0	7	14	0	0
54.5	0	0	0	0	2	10	0	0

[a]All data are CFU/cm surface sampled.

TABLE 14.2 Analysis of Variance of Data Illustrated in Table 14.1[a]

Source of Variation	DF	SS	MS	F	P
Bacterial strain	1	909298	909298	140.81	<0.001
Type of surface	1	719911	719911	111.48	<0.001
Bacteria × surface	1	52480	52480	8.13	<0.05
Main-plot error	4	25480	6457.6		
Time	9	146528276	16280919	5707.7	<0.001
Time × bacteria	9	4332765	481418	168.7	<0.001
Time × surface	9	1215098	135011	47.33	<0.001
3-factor interaction	9	65389	7265	2.54	<0.05
Subplot error	36	102687	2852		

[a]DF = degrees of freedom, SS = sums of squares, MS = mean square, F = variance ratio, P = probability.

interesting, however, is the interaction between type of surface and bacterial species ($F = 8.13$, $P < 0.05$), indicating that differences between surfaces varied with species, there being a slightly greater difference in numbers of *S. epidermidis* that survived between the control and £5 notes. The main effect of time ($F = 5{,}707.7$, $P < 0.001$) reflects the rapid decline in numbers over the period of the experiment in the treatments as a whole, regardless of species or surface. This decline, however, varied with type of surface ($F = 47.33$, $P < 0.001$) and with bacterial strain ($F = 168.7$, $P < 0.001$), a more marked decline being observed on glass coverslips compared with £5 notes and in *E. coli* compared with *S. epidermidis*. There is also a significant three-factor interaction that suggests a more complex relationship between all three variables. In more complex factorial experiments, interactions involving three or more factors may not be easily interpretable; the main effects and two-factor interactions providing most of the useful information. An interesting feature of this design is that despite only two replicates per treatment combination, the subplot error had 36 DF, which should provide sufficient power for testing interactions with time, the principle objective of the experiment. This arises because of the factorial nature of the design and *internal replication* (see Statnote 12).

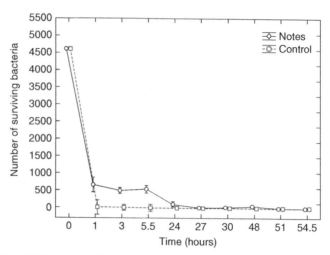

<u>Figure 14.1.</u> Graph illustrating the two factor interaction between type of surface (control coverslips, £5 notes) and sampling time ($F = 47.33$, $P < 0.001$) averaged over bacterial strains. Error bars are 95% confidence intervals.

14.4.3   Repeated-Measures Design and Post Hoc Tests

Repeated measurements made on a single £5 note or coverslip will be highly correlated, and, therefore, the usual post hoc tests cannot be recommended (Snedecor & Cochran, 1980). Nevertheless, a significant interaction between the main-plot factor and time indicates that the *response curve* may vary at different levels of the main-plot factor. A useful first step, therefore, is to examine the graphs of the different effects, an option usually provided by the statistical software. Figure 14.1 illustrates the surface × time interaction effect ($F = 47.33$; $P < 0.001$) and shows a more rapid and pronounced decline in bacterial numbers on coverslips compared with £5 notes. In addition, it may also be possible to partition the main effects and interaction sums of squares into *contrasts* associated with particular types of response curve and test each against the appropriate error (Ridgman, 1975; Snedecor & Cochran, 1980).

14.5   CONCLUSION

Experiments combining different groups or factors and that use ANOVA are a powerful method of investigation in microbiology. ANOVA enables not only the effect of individual factors to be estimated but also their interactions, information that cannot be obtained readily when factors are investigated separately. It is important to consider the design of each experiment because this determines the appropriate ANOVA. Hence, it is necessary to be able to identify the different forms of ANOVA appropriate to different experimental designs and to recognize when a design is fully randomized, in randomized blocks, a split-plot, or incorporates a repeated measure. If there is any doubt about which ANOVA to use in a specific circumstance, the researcher should seek advice from a statistician with experience of research in applied microbiology.

Statnote 15

CORRELATION OF TWO VARIABLES

Nomenclature for variables (X,Y).

Dependent and independent variables.

Pearson's product moment correlation coefficient (r).

Interpretation of r.

Limitations of r.

15.1 INTRODUCTION

Testing the degree of *correlation* between two variables is one of the most commonly used of all statistical methods. Nevertheless, tests based on correlation can be easily misinterpreted, resulting in erroneous conclusions being drawn from an investigation. In the field of microbiology, correlation methods may be used in a wide variety of circumstances. An investigator may wish to study, for example, the correlation between bacterial biomass in a fermentation flask and the concentration of a media supplement, or between the degree of penetration of an antiseptic compound into the skin and skin depth.

There may be several objectives in mind when studying the correlation between two variables. First, an investigator may wish to determine simply whether there is a statistically significant relationship between the two variables, that is, does one variable change in a consistent manner as the other changes? Second, a test of whether the relationship is positive or negative may be required, that is, does one variable increase or decrease as the

Statistical Analysis in Microbiology: Statnotes, Edited by Richard A. Armstrong and Anthony C. Hilton
Copyright © 2010 John Wiley & Sons, Inc.

other changes? Third, a measure of the degree of statistical significance that can be attached to the correlation may be important. Finally, it may be necessary to determine what proportion of the variability in one of the variables can be accounted for or "explained" by the other variable, for example, how much of the variation in bacterial biomass can be explained by the concentration of a media supplement? This statnote examines the use of the most widely used statistic in correlation studies, namely, Pearson's product moment correlation coefficient (r).

15.2 NAMING VARIABLES

A test of correlation establishes whether there is a *linear* relationship between two different variables. The two variables are usually designated as Y the dependent, outcome, or response variable and X the independent, predictor, or explanatory variable. For example, in a study of the relationship between bacterial biomass and concentration of media supplement, the supplement is the independent variable X and bacterial biomass the dependent variable Y, since Y is clearly dependent on X and not vice versa. In some circumstances, it may not be obvious which is the dependent and independent variable, for example, an investigator may wish to study the correlation between two independent measures of the abundance of a bacterium in several soil samples, and the variables should be designated X_1 and X_2. In this and all subsequent statnotes concerned with correlation and regression, uppercase letters are used to label variables and lowercase letters to indicate individual observations.

15.3 SCENARIO

Adequate skin antisepsis prior to invasive procedures is important in preventing infections. Nevertheless, skin antiseptics permeate poorly into the deeper layers of the skin and into hair follicles, which may harbor microorganisms and cause infection when the protective skin barrier is broken. One potential mechanism of delivering antiseptics deeper into the skin is to co-administer a *carrier* to facilitate movement through the various skin layers. Hence, the aim of the study was to evaluate the permeation of a commonly used biocide into the full thickness of human skin when applied alone or in combination with a carrier compound.

Full-thickness human skin samples were obtained from patients undergoing breast reduction surgery. The skin permeation studies were performed with vertical diffusion cells, the stratum corneum of the skin sample being uppermost. One milliliter of antiseptic solution, in the presence and absence of the carrier compound, was aliquoted onto the skin, and incubated for 2, 30, or 24 hours. The assay was performed in triplicate. Following the exposure to the antiseptic solution (+/− carrier) the skin was washed with phosphate-buffered saline (PBS) and three 7-mm punch biopsies taken from each sample. The biopsies were cut with a microtome into 20-μm slices from the skin surface to a depth of 600 μm and 30-μm slices from 600 to 1500 μm. The weight of the skin samples was determined and each analyzed by high-performance liquid chromatography (HPLC) to determine the concentration of antiseptic present as microgram antiseptic per milligram of tissue.

A number of mathematical models might describe the pattern of penetration of the antiseptic. Concentration of the antiseptic appears to decline markedly with skin depth.

Hence, in this preliminary analysis, we wished initially to test whether the antiseptic alone, that is, without carrier and after 30 minutes, penetrated the skin according to a *power law* model, which often describes a rapid decline with distance. A variable Y is distributed as a power law function of X if the dependent variable has an exponent a, that is, a function of the form $Y = cX^{-a}$. If penetration of the antiseptic does follow such a law, then a log–log plot of the data should reveal a linear relationship between concentration of the biocide and skin depth penetration. Hence, log concentration is plotted against log skin depth, and r can be used to test the degree of linearity of the plot.

15.4 DATA

The data comprise several pairs of measurements (x, y) of two variables, namely, concentration of antiseptic (Y) and skin depth (X).

15.5 ANALYSIS

15.5.1 How is the Analysis Carried Out?

The *product moment correlation coefficient* (r) was first proposed by the statistician Karl Pearson in 1902 (Pearson & Lee, 1902). Pearson's r is given by the ratio:

$$r = \frac{\sum xy}{\sqrt{\left(\sum x^2 \sum y^2\right)}} \tag{15.1}$$

where $\sum x^2$ is the SS of the X values, $\sum y^2$ is the SS of the Y values, and $\sum xy$ is the sum of products of the individual pairs of X and Y values, that is, each x is multiplied by the corresponding y and the products summed. Hence, the denominator of r is the square root of the product of the SS of X and Y, that is, it is a combined estimate of the individual variations of the X and Y variables. The numerator of r is a measure of how the two variables vary together. If the individual variations in X and Y were completely explained by the fact that X and Y were linearly related to each other, this ratio would be unity (Snedecor & Cochran, 1980).

15.5.2 Interpretation

The correlation coefficient takes a value from $+1$ to -1 (Fig. 15.1). When $r = +1$, all the data points will lie without deviation on a straight line of positive slope (maximum positive correlation); and, when $r = -1$, all the data points will lie on a straight line of negative slope (maximum negative correlation). By contrast, when $r = 0$, no linearity is present, and the data points are scattered more or less randomly within the two-dimensional space defined by X and Y. Intermediate values of r result from data points scattered around a fitted line—less scatter when r is close to 1 and a greater degree of scatter when r is closer to zero.

Having calculated r from the data, its absolute value, ignoring the sign, is compared to the distribution of r to obtain a P value. This can be obtained from statistical software or by using a table of the correlation coefficient (Fisher & Yates, 1963), entering the table for $n - 2$ DF where n is the number of pairs of observations. Pearson's correlation coefficient has $n - 2$ DF because the mean of both the X and Y values are calculated from

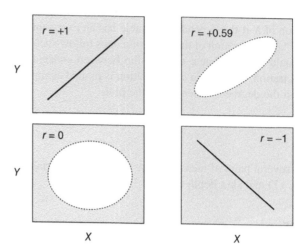

Figure 15.1. Interpretation of Pearson's product moment correlation coefficient (r). When $r = +1$, all the data points will lie on a straight line of positive slope and when $r = -1$, all the data points will lie on a straight line of negative slope. By contrast, when $r = 0$, no linearity is present. Intermediate values of r result from data points scattered around a fitted line; less scatter when r is close to 1 and a greater degree of scatter when r is closer to zero.

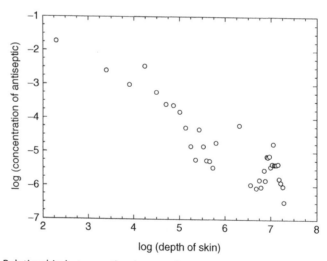

Figure 15.2. Relationship between the degree of penetration of the biocide after 30 minutes (without carrier) and depth of skin (Pearson's correlation coefficient $r = -0.89$, $r^2 = 0.79$).

the data (Snedecor & Cochran, 1980) and, therefore, there are two restrictions in the calculation of r.

The relationship between log concentration of antiseptic and log skin depth is shown in Figure 15.2. The value of r is -0.89, which is highly significant at the $P = 0.001$ level of probability, that is, the probability is less than 1 in 1000 of obtaining a value of *r* of this magnitude by chance, when there is no correlation present. Hence, there is a statistically significant linear decrease in the log of concentration of the antiseptic with the log of increasing skin depth. Hence, penetration of the antiseptic without a carrier is consistent with a power law function, that is, most absorption occurs within the immediate surface layers with very small amounts reaching the deeper layers.

15.6 LIMITATIONS OF *r*

The correlation coefficient *r* has a number of limitations. First, in most applications, *r* tests whether there is a linear relationship between the two variables. If there is a considerable degree of nonlinearity present, a nonsignificant *r* could result even though there is in reality a more complex relationship between *X* and *Y*.

Second, the square of the correlation coefficient r^2 also known as the *coefficient of determination*, measures the proportion of the variance associated with the *Y* variable that can be accounted for or "explained" by the *X* variable. When large numbers of pairs of observations are present, for example, $N > 50$ pairs, examination of the statistical table for *r* (Fisher & Yates, 1963) reveals that values as low as 0.3 could be significant. Hence if $r = 0.3$, then $r^2 = 0.09$ or expressed as a percentage, only 9% of the variation in the *Y* values would be accounted for by the independent variable *X*. This property of *r* can cause considerable confusion in the interpretation of the results of a study. For example, there may be a significant correlation between two variables, but the value of *r* may be so low that the *X* variable accounts for a very small proportion of the variance in *Y*. Hence, a significant *r* would not be biologically meaningful if it accounts for a very small proportion of the variance. In the present example, however, approximately 79% of the variance in log concentration of the biocide was attributable to log of skin depth, a highly significant proportion.

Third, care is needed to ensure that nonhomogeneous groups are not included in the correlation. For example, if measurements of *X* and *Y* were made from two samples of skin differing substantially in depth (say the "surface" and "deeper" layers), then *r* calculated within each group separately may not be significant. It would be inappropriate in this circumstance to combine the groups and to calculate *r* on the pooled data because, even if *r* was then significant, it would not reflect a true linear relationship between *X* and *Y* but only that there was a significant mean difference between the layers.

Fourth, in many correlation studies a significant value of *r* does not imply that there is a "causal" relationship between the two variables or, indeed, that there is any relationship between them at all. Two variables can be mutually correlated not because they are directly related but because both have a significant degree of correlation with a third variable. This circumstance commonly arises in nonexperimental studies where there is a high probability that other variables are involved. There is in fact a strong positive correlation between the consumption of ice cream and the incidence of sunburn! A useful statistic when there are multiple variables present is the *partial correlation coefficient*, that is, the degree of correlation between two variables with the effect of a third *confounding* variable removed (Snedecor & Cochran, 1980). In addition, investigators often use a technique called *stepwise multiple regression* to separate out the effects of individual *X* variables, and this method will be discussed in Statnotes 25 and 26.

15.7 CONCLUSION

Pearson's correlation coefficient (r) is one of the most widely used of all statistics. Nevertheless, care needs to be used in interpreting the results because with large numbers of observations, quite small values of r become significant and the X variable may only account for a small proportion of the variance in Y. Hence, r^2 should always be calculated and included in a discussion of the significance of r. The use of r also assumes that the data follow a bivariate normal distribution (see Statnote 17), and this assumption should be examined prior to the study. If the data do not conform to such a distribution, the use of a nonparametric correlation coefficient should be considered (see Statnote 17). A significant correlation should not be interpreted as indicating "causation" especially in observational studies in which the two variables may be correlated because of their mutual correlations with other confounding variables.

Statnote 16

LIMITS OF AGREEMENT

Agreement and calibration.

Agreement and repeatability.

Measuring agreement: The Bland and Altman plot, bias, and limits of agreement.

16.1 INTRODUCTION

It may be necessary to compare two different methods to estimate the same quantity, and the question may arise as to what extent do the methods agree or disagree with each other. For example, in medicine, a certain measurement may be very difficult to make on a patient without adverse effects so that its true value is unknown. Instead, a new *indirect* method may be used to estimate the measurement, and it is then often necessary to evaluate the new method against the old. In addition, different methods might be used to estimate the numbers of bacteria in a culture and their level of agreement would be important in deciding whether the two methods could be used interchangeably. Note that studies of *agreement* are different from those of *calibration* (see Statnote 19) as in the latter, known quantities are measured by a new method and compared with a highly accurate method to obtain a calibration curve.

Statistical Analysis in Microbiology: Statnotes, Edited by Richard A. Armstrong and Anthony C. Hilton
Copyright © 2010 John Wiley & Sons, Inc.

16.2 SCENARIO

New bacteriological culture media often become available and promise enhancements over existing formulations. Hence, a researcher may wish to compare the performance of the new product against an existing "gold standard" used in a laboratory before adopting the new product into a protocol. If a new product is to be suitable, it would need to perform at least as well as the gold standard in terms of the level of agreement and repeatability of the data generated using the two methods while offering some additional benefit, such as ease of preparation, time saving, or value for money. In this scenario, a substitute product for nutrient agar, namely, Magi-plate, has recently become available on the market. This novel product is offered as a dehydrated powder that, when reconstituted with distilled water, triggers an exothermic reaction of sufficient energy to sterilize the media without autoclaving. According to the claims of the manufacturer, the molten agar is simply poured into sterile Petri dishes and when set offers a culture medium with an identical application range as nutrient agar. This could be an attractive feature for environmental microbiologists on field studies, for example, where access to autoclaving facilities may be limited.

In an initial trial, the agreement and repeatability of nutrient agar and Magi-plate to estimate the viable count of *Escherichia coli* in a broth culture was investigated. Twelve flasks containing 10 ml of sterile nutrient broth were inoculated with one colony of *E. coli* and incubated with shaking for 6 hours at 37°C. Following incubation, serial 10-fold dilutions were prepared from each flask in fresh nutrient broth and the surface of a nutrient agar plate and a Magi-plate inoculated in duplicate (plate A and B) with 0.1 ml of each dilution. In each case the inoculum was spread over the surface of the media using a sterile spreader and the plates incubated at 37°C for 24 hours. Following incubation, the dilution bearing the largest countable number of colonies was selected for the estimation of the viable count in the flask. Counts were corrected for the volume and dilution plated and the CFU per milliliter of the original culture calculated for replicate plates A and B for each culture media.

16.3 DATA

The data comprise two separate estimates (A, B) of the counts of bacteria in 12 separate cultures by each of the two methods (nutrient agar, Magi-plate) and are presented in Table 16.1.

16.4 ANALYSIS

16.4.1 Theory

One method that could be used to assess agreement between two sets of data is to calculate Pearson's correlation coefficient r (see Statnote 15). As shown by Bland and Altman (1986, 1996), however, r is a measure of the *strength* of the relationship between two variables and not the degree of *agreement* between them. A perfect correlation would be present if the points lay along any straight line. However, only if the points lay along the line of equality (the 45° line) would they indicate agreement. Moreover, a highly significant correlation between two variables can hide a considerable lack of agreement. A better

TABLE 16.1   Viable Counts (10^6 per ml) of Bacteria from 12 Samples Determined by Two Methods (Nutrient Agar, Magi-plate) Each on Two Occasions (A, B)

	Method			
	Nutrient Agar		Magi-plate	
Culture Vessel	A	B	A	B
1	1.55	1.54	1.52	1.58
2	1.6	1.61	1.4	1.4
3	1.52	1.66	1.36	1.42
4	1.79	1.45	1.58	1.39
5	1.2	1.12	0.95	1.02
6	1.47	1.7	1.21	1.24
7	1.39	1.58	1.27	1.32
8	1.84	1.78	1.47	1.57
9	1.86	1.8	1.52	1.44
10	1.29	1.37	1.01	1.19
11	1.99	1.89	1.81	1.93
12	1.5	1.49	1.43	1.63

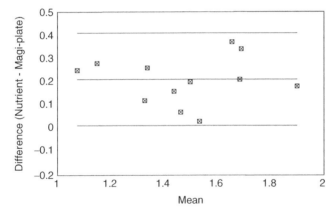

Figure 16.1. Bland and Altman plot of the degree of agreement between bacterial counts obtained by the nutrient agar and Magi-plate methods. The central line is the bias line and the outer lines the 95% confidence intervals.

measure of agreement is to consider by how much does one method differ from the other and how far apart can the measurements be before causing problems. These questions can be answered by constructing a Bland and Altman plot of the data (Bland & Altman, 1986).

16.4.2   How is the Analysis Carried Out?

The essential feature of a Bland and Altman plot is that for each pair of values the difference between them is plotted against the mean of the two values. A typical plot is shown in Fig. 16.1. The mean of all pairs of differences is known as the degree of *bias* and is the central line plotted on the figure. Either side of the bias line is the 95% confidence intervals in which it would be expected that 95% of the differences between the two

TABLE 16.2   Summary of the Bland and Altman Analyses to Test the Degree of Agreement within Both Methods (Comparing A with B) and to Compare the A Data for Both Methods[a]

Comparison	Bias	SD	Limits of Agreement	Lower Limit	Upper Limit
Nutrient agar (A with B)	0	0.15	0.30	−0.30	+0.30
Magi-plate (A with B)	0.05	0.11	0.22	−0.17	+0.27
Nutrient agar with Magi-plate (A data only)	0.21	0.10	0.20	−0.01	+0.41

[a] SD = standard deviation.

methods would fall. In the present scenario, the Bland and Altman analysis was used, first, to test the degree of agreement within both methods (comparing A with B within a method) and, second, to compare the A data only using both methods.

16.4.3   Interpretation

The Bland and Altman analyses are summarized in Table 16.2. There is a good agreement within methods, that is, comparing A with B within each method separately. For the nutrient agar the degree of bias was actually zero and 95% of differences between A and B would fall between −0.30 and +0.30. For the Magi-plate data, the degree of bias is 0.05 (95% confidence intervals −0.17 to +0.27). The degree of disagreement, however, was greater between the two methods (Fig. 16.1); the degree of bias being +0.21, that is, the Magi-plate data gave, on average, scores of 0.21 less than the nutrient agar (95% confidence intervals −0.01 to +0.41). If relevant, the degree of bias can be used to "correct" the counts obtained by one method as if they had been obtained by the other.

16.5   CONCLUSION

If a quantity has been measured by two different methods, the degree of agreement between them should not be tested using Pearson's correlation coefficient r. Instead the differences between the two methods should be compared with their mean difference using a Bland and Altman plot. Such a plot illustrates the level of agreement between two methods and enables the degree of bias of one method over the other to be calculated and applied if necessary as a correction factor.

Statnote 17

NONPARAMETRIC CORRELATION COEFFICIENTS

Bivariate normal distribution.
Spearman's rank correlation.
Kendall's rank correlation.
Gamma.

17.1 INTRODUCTION

The most common method of testing the degree of correlation between two variables (X, Y) is to use Pearson's correlation coefficient r (see Statnote 15). Although Pearson's r is widely used with all types of data, it is essentially a parametric test and is based on the assumption that the individual pairs of values (x, y) are members of a *bivariate normal distribution* (Snedecor & Cochran, 1980). If the data depart significantly from such a distribution, then a nonparametric correlation coefficient may be more appropriate. This statnote describes the most widely used of the nonparametric correlation coefficients, namely Spearman's rank correlation (ρ or r_s), Kendall's tau (τ), and gamma (Snedecor & Cochran, 1980).

17.2 BIVARIATE NORMAL DISTRIBUTION

Pearsons' r ssumes that the data follow, at least approximately, a bivariate normal distribution. The bivariate normal distribution is an extension of the normal distribution from

Statistical Analysis in Microbiology: Statnotes, Edited by Richard A. Armstrong and Anthony C. Hilton
Copyright © 2010 John Wiley & Sons, Inc.

one to two variables and has the following properties: (1) for each value of X, the corresponding values of Y are normally distributed, the means of these normal distributions lying on a straight line and the variance being constant for each x; (2) for each Y, the corresponding X values are normally distributed; and (3) the marginal distributions of X and Y are also normally distributed. This distribution is, therefore, defined by five parameters (the means and standard deviations of the X and Y variables and the population correlation coefficient p). The data are likely to approximate to such a distribution if both X and Y are continuous variables and themselves normally distributed. Some statisticians have argued that if a test of the H_0 in which there is no correlation between X and Y is required, Pearson's r may still be used providing one of the variables is normally distributed (Snedecor & Cochran, 1980). If one or both variables are small whole numbers, scores are based on a limited scale, or percentages; however, a nonparametric correlation coefficient should be considered as an alternative.

17.3   SCENARIO

Many species of bacteria and fungi have been recorded in the proximity of plant roots, a region known as the *rhizosphere*, and some species may be exclusively rhizosphere organisms (Burges, 1967). The root surface offers many potential sources of food for soil microorganisms as plant roots exude substances into the soil, the type and amount depending on plant growth conditions. In addition, many of the organisms that accumulate on the root surface may be deficient in one or more substances such as carbohydrates, amino acids, thiamine, biotin, organic acids, nucleotides, or various enzymes, and many of these substances can only be obtained from the rhizosphere. The degree of root secretion is often dependent on the type of plant, for example, legumes have high rates of root secretion while cruciferous plants tend to exude less. An investigator wished to test the hypothesis that the general abundance of fungi close to the root surface was correlated with the degree of carbohydrate secreted from the root.

Ten plants of the legume *Trifolium pratense* L., obtained from cuttings of wild plants, were propagated each in its own pot in garden compost for a period of 6 months. After 6 months, a 1-g soil sample was taken from the rhizosphere of each pot at a depth of 3 cm. Carbohydrate analysis was carried out on these samples using gas–liquid chromatography (GLC), 5 replicate analyses being made from each sample. In addition, 10 further small soil samples (5 to 15 mg) were taken at different depths adjacent to the plant roots using the flattened blade of a sterilized nichrome inoculating needle, and used to crush and disperse the soil aggregates in the bottom of each of 10 sterile Petri dishes. A little sterile water was then added to assist the dispersion. Melted and cooled agar (8 to 10 ml) was poured into the dish and manipulated before setting so as to secure as complete a dispersion of the soil sample as possible. After incubation, the general abundance of fungal colonies on each dish was expressed on a ranked scale: 0 (none), 1 (few), 2 (frequent), or 3 (abundant). The data were averaged over the 10 samples collected from each plant.

17.4   DATA

The data comprise two variables (X, Y), one of which is a score of fungal abundance (Y) and the other a measurement of carbohydrate concentration (X) measured on each of 10 plants. The total carbohydrate levels in the soil samples are continuous data, measured to four significant figures, and can be regarded as normally distributed. The abundance of the fungi, however, is expressed on a ranked scale. These data are less likely to be normally distributed, and, therefore, there may be doubt as to whether the data as a whole conform to the bivariate normal distribution. Pearson's r could still be used in this circumstance, as one variable is normally distributed, but a better approach might be to use a nonparametric correlation coefficient.

17.5 ANALYSIS: SPEARMAN'S RANK CORRELATION (ρ, R_S)

17.5.1 How Is the Analysis Carried Out?

The calculations involved in making a Spearman rank correlation test (Spearman, 1904) on an actual data set are shown in Table 17.1. Essentially, ranks are assigned to the X and Y values separately within each column, starting with the lowest value and ending with the highest, a procedure that was also employed in Statnote 4. If some values within the column are the same (called *ties*), they get the mean of the ranks that would have been assigned to these values if they had been different. The ranks are subtracted for each pair of values ($D = X - Y$), the differences are squared (D^2), and the sum of the

TABLE 17.1 Measuring Degree of Correlation between General Abundance of Fungi in the Rhizosphere (Average of Five Plates Scored on Four-Point Scale) (Y) and Degree of Secretion of Carbohydrate (μg mg dry weight of soil) (X) Using Spearman's Rank Correlation (r_s)

	X	Y	Rank X	Rank Y	$X - Y$	D^2
1	51.97	3.4	7	9	-2	4
2	22.23	2.1	3	5	-2	4
3	17.81	1.8	2	2	0	0
4	60.54	2.9	8	6	2	4
5	82.31	3.6	10	10	0	0
6	24.60	1.9	4	3	1	1
7	39.71	3.0	6	7	-1	1
8	61.23	3.2	9	8	1	1
9	14.10	1.1	1	1	0	0
10	29.80	2.0	5	4	1	1

1. The ranks are subtracted for each pair of values ($D = X - Y$).
2. Square the differences (D^2) and add up the squared ranks.
3. Calculate ($\Sigma D^2 = 16$).
4. Calculate $N^* = n^3 - n = 990$ where n is the number of pairs of observations.
5. If there are no ties, as in the present example, Spearman's rank correlation (r_s) is given by
   $r_s = 1 - (6 \times \Sigma D^2 / N^*)$.
6. It ties are present, calculate $\Sigma(T^3 - T)$ for each variable where T_Y and T_X are the number of ties in each run of Y and X, respectively. Then calculate $SS_Y = (N^* - T_Y)/12$ and
   $SS_X = (N^* - T_X)/12$ and $r_s = (SS_Y + SS_X - \Sigma D_2)/2 \times \sqrt{(SS_Y \times SS_X)}$.
7. If the number of pairs is <10, then a table of Spearman's rank correlation can be used to obtain a P value (Snedecor & Cochran, 1980). If $N > 10$, then the table of Pearson's r can be used to test significance (Fisher & Yates, 1963).
8. Spearman's correlation coefficient (r_s) = 0.903 ($P < 0.01$).

squared ranks calculated ($\Sigma D^2 = 16$). The value of N^* is then calculated. If there are no ties, as in the present example, calculation of Spearman's rank correlation (r_s) is relatively straightforward; but, if ties are present, the calculation is more complex.

17.5.2   Interpretation

In the present case, the value of Spearman's correlation coefficient (r_s) was 0.903 and $P < 0.01$, indicating a significant positive correlation between the general abundance of fungi in the rhizosphere of *T. pratense* and the degree of secretion of carbohydrate from the plant roots.

17.6   ANALYSIS: KENDALL'S RANK CORRELATION (τ)

Kendall's rank correlation (τ) (Kendall, 1938) is another nonparametric method of testing the degree of correlation between two variables and like r_s can be used as a measure of ability to appraise or detect a property by scoring. Like r_s, τ varies from +1 (complete concordance) to −1 (complete disagreement), but it is calculated differently (Snedecor & Cochran, 1980). Kendall's rank correlation is closely related to Spearman's r_s, and it probably matters little in most applications which method is actually used. One advantage of τ is that it can be extended to study *partial correlations*, a statistical method that will be discussed in Statnote 25.

17.7   ANALYSIS: GAMMA (γ)

A third type of nonparametric correlation coefficient often available in statistical software is called *gamma*. Gamma is probably closer to Kendall's τ than Spearman's r_s but is regarded as a preferable test if the data contain many tied values.

17.8   CONCLUSION

If one or both variables in a correlation test are small whole numbers, scores based on a limited scale, or percentages, a nonparametric correlation coefficient should be considered as an alternative to Pearson's r. Kendall's τ and Spearman's r_s are similar, but the former should be considered if the analysis is to be extended to include partial correlations. If the data contain many tied values, then gamma should be considered as an alternative.

Statnote 18

FITTING A REGRESSION LINE TO DATA

Fitting a straight line to data.

Assumptions of linear regression.

Goodness of fit of the line to the data points: coefficient of determination (r^2), analysis of variance (ANOVA), and t test of the slope of the line.

18.1 INTRODUCTION

In Statnotes 15 and 17, the use of correlation methods to analyze the relationship between two variables was described. One of the most widely used statistics is Pearson's correlation coefficient r, which tests whether there is a linear correlation between two variables X and Y. Once a linear correlation between two variables has been established, however, a regression line can be fitted to the data to describe the relationship between the two variables in more detail.

Regression analysis is one of the most useful statistical methods in microbiology and has many uses. For example, the objective may be to determine simply whether a relationship exists between Y and X, to study the shape of the relationship (whether linear or curved), to establish a mathematical equation linking Y and X, or to predict the value of Y for a new value of X. This statnote describes the methods used to fit a regression line to data and how to test the statistical significance of the line, that is, how good a fit the line is to the data points. Subsequent statnotes will describe how to use the line for

Statistical Analysis in Microbiology: Statnotes, Edited by Richard A. Armstrong and Anthony C. Hilton
Copyright © 2010 John Wiley & Sons, Inc.

prediction and calibration (see Statnote 19), how to compare two or more regression lines (see Statnote 20), and how to fit selected nonlinear types of regression (see Statnotes 21 to 23).

18.2 LINE OF BEST FIT

A study was carried out to investigate whether bacterial biomass in a fermentation flask (Y) was linearly related to the concentration of media supplement supplied (X). Eight flasks each with a different concentration of medium were inoculated with a bacterium. Hence, bacterial biomass is considered to be dependent upon media concentration. Hence, Y is the *dependent*, *outcome*, or *response* variable and X the *independent*, *predictor*, or *explanatory* variable.

The first stage in a regression analysis is to plot the Y and X values on a graph (Fig. 18.1), and it is obvious that there is an increase in biomass with increasing concentration of media. The straight line that provides the best fit to these data is called the *sample regression* of Y on X or the *fitted line*. This regression line has the following equation:

$$Y = a + bX \qquad\qquad (18.1)$$

where a is the point at which the line cuts or "intercepts" the Y axis and b is the slope of the line, that is, the rate of change in Y per unit of X. This equation may look familiar and is indeed derived in the same way as the straight line $Y = MX + C$ more frequently encountered in a mathematical context.

There are various methods for calculating the line of best fit, the most widely used being the *least squares method*. First, the line passes through the center of the cluster of points, that is, the point with coordinates that are the means of the X and Y values (x^*, y^*). The distance d measures the difference on the Y scale between the actual value

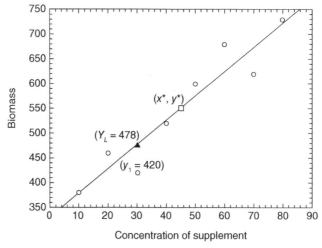

Figure 18.1. Fitting regression line to data. The point (x^*, y^*) represents the mean of X and Y values. Distance d measures difference on Y scale between actual value of Y at that point (y_1) and the corresponding value of Y on the fitted line (y_L), i.e., $d = y_1 - y_L$.

of Y at that point (y_1) and the corresponding value of Y on the fitted line (y_L), that is, $d = y_1 - y_L$. Hence, the distances d, calculated for all data points, measure the deviations of the points from the fitted line. The fitted regression line satisfies two further conditions: (1) that the sum of the distances d from the line is zero ($\Sigma d = 0$), that is, approximately half the data points will be above and half below the line and (2) that the SS of the distances (Σd^2) will be as small as possible, that is, the SS of the deviations from the line (hence, least squares) is minimized. It can be shown for N pairs of observations that there is only one straight line that can be drawn that fulfills all of these conditions, and this line has a slope b given by the following equation:

$$b = \frac{\sum xy - \left(\sum x \sum y / n\right)}{\sum x^2 - \left(\sum x\right)^2 / n} \qquad (18.2)$$

The equation for the slope of the line b is similar to that of Pearson's r (see Statnote 15). The numerator is the sum of products, that is, the sum of the x and y values multiplied together, but the denominator is the SS of the x values alone. The slope b, also known as the *regression coefficient*, is an estimate of the average change in Y associated with a unit increase in X. It is not intuitively obvious why Eq. (18.2) should be an estimate of the slope of the regression line b. This equation, however, is based on two independent estimates of the increase in Y per unit of X and is actually the weighted mean of these estimates (Snedecor & Cochran, 1980).

18.3 SCENARIO

We return to the scenario first described in Statnote 15. Essentially, adequate skin antisepsis prior to invasive procedures is important in preventing infections. Nevertheless, skin antiseptics permeate poorly into the deeper layers of the skin and into hair follicles, which may harbor microorganisms and cause infection when the protective skin barrier is broken. One potential mechanism of delivering antiseptics deeper into the skin is to co-administer a *carrier* compound to facilitate movement of the biocide through the various skin layers. Hence, the aim of the study was to evaluate the permeation of a commonly used biocide into the full thickness of human skin when applied alone or in combination with a carrier compound.

Full-thickness human skin samples were obtained from patients undergoing breast reduction surgery. The skin permeation studies were performed with vertical diffusion cells with the stratum corneum of the skin sample uppermost. One milliliter of antiseptic solution in the presence or absence of the carrier compound was aliquoted onto the skin and incubated for 2 minutes, 30 minutes, or 24 hours. The assay was performed in triplicate. Following the exposure to the antiseptic solution (+/− carrier), the skin was washed with PBS and three 7-mm punch biopsies taken from each sample. The biopsies were cut with a microtome into 20-µm slices from the skin surface to a depth of 600 µm and 30-µm slices from 600 to 1500 µm. The weight of the skin samples was determined and each analyzed by HPLC to determine the concentration of antiseptic present as microgram antiseptic per milligram of tissue. A number of mathematical models might describe the pattern of penetration of the antiseptic into the skin. In the correlation analysis described in Statnote 15, we tested a specific model by which the antiseptic penetrated into the thickness of the skin, namely whether the antiseptic alone, that is, without carrier and after

30 minutes, would have penetrated the skin according to a power law model. A variable Y is distributed as a power law function of X if the dependent variable has an exponent a, that is, a function of the form $Y = CX^{-a}$. If penetration of the antiseptic does follow such a law, then a log–log plot of the data should be linear.

18.4   DATA

The data comprise several pairs of measurements of two variables, namely, the concentration of antiseptic (Y) and skin depth (X) and the data are presented in graphical form in Figure 18.2.

18.5   ANALYSIS: FITTING THE LINE

A linear regression line has been fitted to the data in Figure 18.2 by the method of least squares and has the following equation:

$$\log Y = 0.0114 - 0.811\log X \tag{18.3}$$

The regression analysis confirms the negative relationship between concentration of biocide and skin depth and provides a mathematical equation that describes the relationship between these two variables.

18.6   ANALYSIS: GOODNESS OF FIT OF THE LINE TO THE POINTS

18.6.1   Coefficient of Determination (r^2)

The square of Pearson's correlation coefficient r^2 (also known as the coefficient of determination) represents the proportion of the variance of the Y values attributable to the

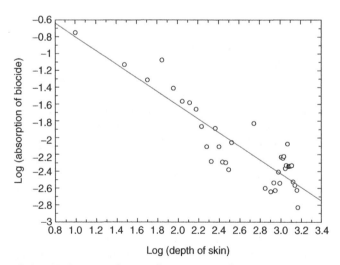

Figure 18.2. Relationship between degree of penetration of biocide (without carrier) and depth of skin (Pearson's correlation coefficient $r = -0.89$, $r^2 = 0.79$; $F = 144.72$, $P < 0.001$; $t = 12.03$).

linear regression on X (see Statnote 15). Hence, r^2 rovides an estimate of the "strength" of the relationship between Y and X. In the present example, $r = -0.89$ and $r^2 = 0.79$, suggesting that 79% of the variance in log concentration of biocide was attributable to log skin depth. Hence, the line "accounts for" or "explains" 79% of the variation in concentration of biocide with depth, a highly significant proportion of the total. Note that there is no established "cut-off" in r^2 below which the line would be regarded as a poor fit. Nevertheless, a line accounting for less than 50% of the variance would be regarded as a poor fit to the data.

18.6.2 Analysis of Variance

Goodness of fit of the line can also be tested using ANOVA (see Statnote 6). ANOVA determines the *statistical significance* of the line rather than the strength of the relationship between the two variables. In an ANOVA, the total variation of the Y values is divided into a *linear effect*, that is, that portion of the variance accounted for by the line, and the *error variance* associated with deviations from the line, the two sources of variation being compared using a variance ratio F test. The total variation is the SS of the deviations of the Y values (Y_{ss}) from their mean (y^*) and is given by

$$Y_{ss} = \sum y^2 - \frac{\left(\sum y\right)^2}{n} \tag{18.4}$$

The total variation can then be partitioned into the linear effect and error. The linear effect is the variation accounted for by the points that lie along the fitted regression line (y_L) from their mean (y^*):

$$\text{Linear effect} = \frac{\left\{\sum xy - \left(\sum x\right)\left(\sum y\right)/n\right\}^2}{\sum x^2 - \left(\sum x\right)^2/n} \tag{18.5}$$

This equation closely resembles that for the slope of the line b [Eq. (18.2)] except that the numerator, the sum of products, is squared. The residual or error variation is based on the deviation of the actual y values from the regression line (i.e., the d values) and can be determined by subtraction:

$$\text{Error variation} = \text{SS of } Y \text{ values} - \text{Linear effect} \tag{18.6}$$

The ANOVA is shown in Table 18.1. The F ratio is, therefore, a comparison of the magnitude of the linear effect with the degree of deviation of the points from the line. Hence, for there to be a statistically significant linear effect, the variation accounted for by the line must be significantly greater than the variation of the data points from the line. To obtain a P value, the value of F can either be taken to a table of the F ratio (Fisher & Yates, 1963; Snedecor & Cochran, 1980), entering the table for 1 and $N - 2$ DF (where N is the number of pairs of observations), or the P value can be obtained from statistical software. If F is greater than the critical value at $P = 0.05$, there is a statistically significant fit to the data points. In the present example, $F = 144.72$ giving a P value considerably less than 0.001. The probability of obtaining an F value of this magnitude by chance is less than 1 in a 1000, and hence the regression line exhibits a highly significant fit to the data.

TABLE 18.1   Analysis of Variance (ANOVA) of Linear Regression of Log Concentration of Biocide (Y) in Relation to Log Skin Depth (X)[a]

Effect	DF	SS	MS	F	P
Linear	1	7.0902	7.0902	144.72	<0.001
Error	36	1.76378	0.048994		

[a] DF = degrees of freedom, SS = sums of squares, MS = mean square, F = variance ratio, P = probability.

18.6.3 t Test of Slope of Regression Line

A third method of judging goodness of fit is to test whether the slope of line b is significantly different from zero. The ratio of b to its standard error (SE) (s_b) converts b so that it is a member of the t distribution. The SE of the slope of the line is given by the following equation:

$$s_b = \sqrt{\frac{\text{Mean square error}}{\sum x^2 - \left(\sum x\right)^2 / n}} \tag{18.7}$$

The mean square error is taken from the ANOVA (Table 18.1). The value of t (b/s_b) is taken to the t table with $N - 1$ DF where N is the number of pairs of observations. In this example, a highly significant value of t was obtained ($t = 12.03$, $P < 0.001$), indicating that the slope of the line is significantly different from zero.

To decide which of the three methods of testing a regression line to use depends on the precise hypothesis posed by the study. Hence, r^2 estimates the strength of the relationship between Y and X, ANOVA whether the regression line is statistically significant, and the t test whether the 'slope of the line is significantly different from zero.

18.7 CONCLUSION

Fitting a linear regression to data provides much more information about the relationship between two variables than a simple correlation test. A goodness-of-fit test of the line should always be carried out. Hence, r^2 estimates the strength of the relationship between Y and X, ANOVA whether a statistically significant line is present, and the t test whether the slope of the line is significantly different from zero. In addition, it is important to check whether the data fit the assumptions for regression analysis and, if not, whether a transformation of the Y and/or X variables is necessary.

Statnote 19

USING A REGRESSION LINE FOR PREDICTION AND CALIBRATION

Prediction and calibration.

Types of prediction problem.

Standard error (SE) of predicted values.

19.1 INTRODUCTION

In Statnote 18, the use of regression methods to analyze the relationship between two variables X and Y was described. The method of least squares was used to fit a regression line to the data, and various methods of testing the goodness of fit of the line to the points were described. A regression analysis can also be used to predict a value of Y from a new reading of X, for example, to predict bacterial cell number from optical density (OD) readings. This aspect of regression studies is also called *calibration*, that is, estimating a quantity that may be difficult to measure from a variable that is easier to measure. This statnote describes the various predictions that can be made using a regression line and discusses their limitations.

19.2 TYPES OF PREDICTION PROBLEM

There are two major types of prediction problem that can be solved using regression, and it is important to understand the differences between them. First, there is the prediction

Statistical Analysis in Microbiology: Statnotes, Edited by Richard A. Armstrong and Anthony C. Hilton

of the *population regression line* μ for a new value of x. Hence, an investigator may wish to make inferences about the *height* of the population regression line at the point x, that is, the average value of Y associated with a value of x. Second, there is prediction of an *individual new member* of the population y_1 for which x_1 has been measured. The second problem is probably the most commonly encountered and the most relevant to calibration studies. In both of these prediction problems, however, the predicted value of Y is actually the same, but the standard errors (SE) associated with the predictions will be different. This is because in the first instance a population value or mean is being estimated while in the second an individual value is being estimated. Formulas for the calculation of the different SE corresponding to these two prediction problems are given by Snedecor and Cochran (1980). In most cases, significantly greater errors will result when estimating an individual rather than a population value.

19.3   SCENARIO

In some experimental protocols it is necessary to estimate the number of bacterial cells present in a culture broth where time limitations may prevent the use of a standard culture-based colony-counting technique. In these circumstances, a researcher may employ direct counting methods using microscopy and a hemocytometer slide, for example, or more frequently use measurements of OD of the culture broth as a prediction of cell number. In this latter situation, a calibration graph must first be derived to reveal the relationship between the CFU per milliliter of culture broth for a given bacterium and the corresponding OD usually measured at a wavelength of 600 nm.

A 10-ml volume of sterile nutrient broth was inoculated with a culture of *Staphylococcus aureus* and incubated at 37°C for 24 hours. Following incubation, the culture was serially diluted down to 10^{-9} by mixing 1 ml of culture with 9 ml of fresh nutrient broth in a sterile Universal tube. From each of the tubes within the prepared dilution series, 1 ml of culture media was transferred into a disposable plastic cuvette and the OD measured at 600 nm using a standard spectrophotometer, which had been previously blanked against a cuvette containing uninoculated nutrient broth. In a similar manner, from each of the tubes within the first dilution series, further serial dilutions in sterile nutrient broth were prepared as required, and 0.1 ml of this second dilution series was inoculated onto the surface of a nutrient agar plate. The inoculum was spread across the surface of the agar using a sterile spreader and the plates incubated at 37°C for 24 hours. Following incubation, the CFU/ml in each of the tubes within the first dilution series was calculated by counting the colonies at an appropriate dilution.

19.4   DATA

The data were collated and are presented graphically in Figure 19.1 by plotting the CFU/ml on the Y axis against the corresponding OD_{600} measurement on the X axis.

19.5   ANALYSIS

19.5.1   Fitting the Regression Line

The first stage in a calibration study is to fit a regression line to the data using the method of least squares and to test its goodness of fit as described in Statnote 18. To estimate bacterial cell numbers from OD requires fitting the regression of bacterial numbers (Y) on OD (X), and this line (Fig. 19.1) has the following equation:

$$Y = -1.333 \times 10^9 + 3.111 \times 10^9 X \qquad (19.1)$$

The goodness of fit tests (see Statnote 18) suggest that the line is a good fit to the data. Hence, $r^2 = 0.98$, that is, 98% of the variance in bacterial numbers is accounted for by OD, the value of $F = 398.0$ suggests a highly significant line is present ($P < 0.001$), and the t test of the slope of the line gave a value of $t = 19.95$, which suggests the line has a highly significant slope ($P < 0.001$). All of these tests provide confidence that the regression line is a good fit to the data and is therefore suitable for calibration.

19.5.2 Confidence Intervals for a Regression Line

Two sets of 95% confidence intervals have been fitted to the regression line in Figure 19.1. The inner confidence bands are the confidence limits for predicting the population regression line. Hence, we would be 95% confident that the average value of Y for any x would lie within these boundaries. The outer confidence intervals are for predicting an individual y corresponding to a value of x. It is important not to confuse the two types of prediction problem. If, for example, the regression of weight on height was plotted for a sample of 20-year-old men (Snedecor & Cochran, 1980), the purpose might be to predict the average weight of such men at a specific height (inner confidence bands) or the individual weight of a new male whose height was known (outer confidence bands). As mentioned earlier, the two estimates are actually the same, but the SE and therefore the confidence bands are different.

Using Figure 19.1 for prediction and calibration, the predicted value of y for a new value x is

$$y = Y^* + bx \qquad (19.2)$$

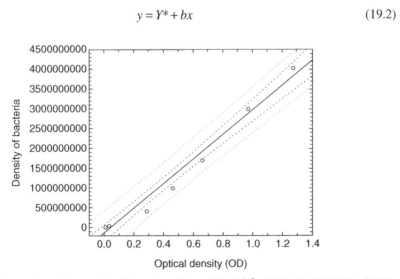

Figure 19.1. Regression of bacterial cell numbers (Y) on OD (X) [$r^2 = 98\%$, $F = 398.04$ ($P < 0.001$), $t = 19.95$ ($P < 0.001$)]. The two sets of confidence bands represent confidence intervals for the population regression line (inner confidence bands) and for making individual predictions of y for a new value x (outer confidence bands).

where $Y*$ is the mean of the Y values and b the slope of the line. Hence, if OD = 1.1, then estimated y is 3.28885×10^9 (95% confidence limits: upper limit = 3.59498×10^9; lower limit = 2.982719×10^9).

19.5.3 Interpretation

There are a number of problems that need to be considered when using a regression line for calibration. First, the confidence bands are usually parabolic in shape (not readily apparent in the present data) and have curved borders that may widen significantly at the limits of the data, especially if the line is a poor fit to the data. Hence, estimates of y can have large errors at the limits of the data. As discussed in Statnote 16, there is no established "*cut-off*" in r^2 below which the line would be regarded as a poor fit for calibration purposes. As the value of r^2 decreases, however, the confidence bands widen and predictions of y become increasingly inaccurate. A regression line is likely to be most useful for calibration: (1) if the range of values of the X variable is large, (2) if there is a good representation of the x,y values across the range of X, and (3) if several estimates of y are made at each x. Second, a regression line is sometimes used several times in the course of an investigation to predict a number of new values. In this circumstance, the probability that all of the confidence intervals include the correct value of Y will be less than $P = 0.95$ and a correction of the P values may be necessary using *Bonferroni's inequalities* (Snedecor & Cochran, 1980). Third, Y may have been measured at several fixed values of X, but the intention may be to predict X from Y, and this prediction must be made from the regression of Y on X. This is a significantly more complex calculation and the method, together with the appropriate SE, is described in Snedecor and Cochran (1980).

19.6 CONCLUSION

Two types of prediction problem can be solved using a regression line, namely, prediction of the population regression line μ at the point x and prediction of an individual new member of the population y_1 for which x_1 has been measured. The second problem is probably the most commonly encountered and the most relevant to calibration studies. A regression line is likely to be most useful for calibration if the range of values of the X variable is large, if there is a good representation of the x, y values across the range of X, and if several estimates of y are made at each x. It is poor statistical practice to use a regression line for calibration or prediction beyond the limits of the data.

Statnote 20

COMPARISON OF REGRESSION LINES

Differences between regression lines.
Analysis of covariance (ANCOVA).
Comparing the slopes and elevations of two regression lines.

20.1 INTRODUCTION

The relationship between two variables (X, Y) may have been studied at various times or in different laboratories, giving rise to two or more independent estimates of the relationship between Y and X. In these circumstances, it may be of interest to discover whether the different regression lines are the same. An investigator may wish to determine the relationship between Y and X given one set of conditions compared to the relationship under a different set of conditions. In addition, if two or more regression lines are the same, an investigator may wish to combine the data from the different studies and fit a single regression line to the data as a whole. This statnote describes the statistical methods associated with comparing two or more regression lines.

20.2 SCENARIO

We return to the scenario first described in Statnote 15. Essentially, adequate skin antisepsis prior to invasive procedures is important in preventing infections. Nevertheless,

Statistical Analysis in Microbiology: Statnotes, Edited by Richard A. Armstrong and Anthony C. Hilton
Copyright © 2010 John Wiley & Sons, Inc.

skin antiseptics permeate poorly into the deeper layers of the skin and into hair follicles, which may harbor microorganisms and cause infection when the protective skin barrier is broken. One potential mechanism of delivering antiseptics deeper into the skin is to co-administer a carrier compound to facilitate movement of the biocide through the various skin layers. Hence, the aim of the study was to evaluate the permeation of a commonly used biocide into the full thickness of human skin when applied alone or in combination with a carrier compound.

Full-thickness human skin samples were obtained from patients undergoing breast reduction surgery. The skin permeation studies were performed with vertical diffusion cells with the stratum corneum of the skin sample uppermost. One milliliter of antiseptic solution in the presence or absence of the carrier compound was aliquoted onto the skin and incubated for 2 minutes, 30 minutes, or 24 hours. The assay was performed in triplicate. Following the exposure to the antiseptic solution (+/− carrier) the skin was washed with PBS and three 7-mm punch biopsies taken from each sample. The biopsies were cut with a microtome into 20-μm slices from the skin surface to a depth of 600 μm and 30-μm slices from 600 to 1500 μm. The weight of the skin samples was determined and each analyzed by HPLC to determine the concentration of antiseptic present as microgram antiseptic per milligram of tissue.

In the correlation analysis described in Statnote 15, we wished to describe and test possible models by which the antiseptic penetrated into the skin. Specifically, we tested whether the antiseptic alone would have penetrated the skin according to a power law model. A variable Y is distributed as a power law function of X if the dependent variable has an exponent a, that is, a function of the form $Y = CX^{-a}$. If penetration of the antiseptic does follow such a law, then a log–log plot of the data should be linear. In Statnote 15, this model was fitted to a single set of data, that is, without the presence of a carrier compound. In this statnote, we extend the analysis to test whether the presence of a carrier compound affected the degree of penetration through the skin.

20.3   DATA

The data comprise two sets of measurements of two variables, namely, the concentration of antiseptic (Y) and skin depth (X) with (Y_1, X_1) and without carrier (Y_2, X_2) and the data are presented graphically in Figure 20.1.

20.4   ANALYSIS

20.4.1   How Is the Analysis Carried Out?

Regression lines may differ in three properties. First, they may differ in *residual variance*, that is, in the degree of scatter of the points about the lines since one line may fit the data better than the other. Second, they may differ in *slope b*, that is, one line may exhibit a greater change in Y per unit of X than the other. Third, the lines may differ in *elevation a*, that is, if the two lines are parallel, they will intersect the Y axis at different points. To test whether the residual variances differ, the larger variance is divided by the smaller to obtain an F value as described in Statnote 8. Whether the two lines differ in slope or elevation can be tested using an extension of ANOVA termed *analysis of covariance* (ANCOVA) (Snedecor & Cochran, 1980). Essentially, individual regression lines are fitted to each set of data separately, and a further common line is fitted to the data pooled from both sets of data; the analysis essentially comparing the individual fits with that of the pooled regression. Differences in the slopes and elevations of the lines can then be tested. If there is a significant difference between the slopes of the regression lines, then it is not necessary to test their elevations as these will vary at each value of X. If the lines have similar slopes, then their elevations can also be tested. If two or more regression lines are shown to have the same slopes and elevations, then the investigator may wish to combine the data in a single regression analysis. It may not be obvious whether or not a particular software package will carry out this analysis. Software such as STATISTICA and SPSS, however, which incorporate a *general linear modeling* option, will usually enable this type of analysis to be performed.

20.4.2 Interpretation

The first stage in the analysis is to fit a linear regression to each set of data separately and the lines of best fit are shown in Figure 20.1. Examination of the regression lines suggests that they differ in slope. The values of r and r^2 indicate that both sets of data can be fitted adequately by straight lines, although the linear regression is a better fit to the data without an added carrier compound. A regression line is then fitted to the data pooled from both data sets, and this enables the slopes and elevations of the lines to be compared. The analysis of variance table is shown in Table 20.1. There is a highly significant difference between the slopes of the two regression lines ($F = 39.63$, $P < 0.001$). The comparison between heights or elevations of the two regressions is also included here but would not be relevant in this example as the lines have different slopes. Note that if the two lines were parallel, then it would be important to test differences in elevation.

Hence, the regression of log absorption of biocide declines less rapidly with log skin depth when a carrier compound is present. Note that the Y axis represents a negative logarithmic scale, and hence the lower negative numbers represent a greater degree of absorption. It is, therefore, concluded that the presence of the carrier does alter the pattern of

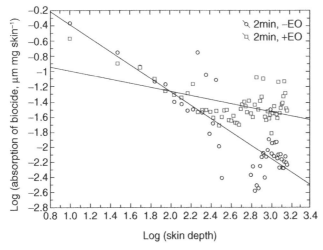

Figure 20.1. Relationship between degree of penetration of biocide with (+EO) and without (−EO) carrier and depth of skin (Pearson's correlation coefficient without carrier $r = -0.84$, $r^2 = 0.71$; with carrier $r = 0.53$, $r^2 = 0.28$).

TABLE 20.1   Analysis of Covariance of Data in Figure 20.1 Comparing the Slopes and Elevations of Two Regression Lines[a]

Variation	DF	SS	MS	F	P
Between slopes	1	2.2375	2.2375	39.63	<0.001
Error	102	5.7596	5.65×10^{-2}		
Between elevations	1	4.4217	4.4217		
Error	103	7.9971	7.7642×10^{-2}		

[a] DF = degrees of freedom, SS = sums of squares, MS = mean square, F = variance ratio, P = probability.

penetration of the antiseptic into the skin and increases the amount of antiseptic reaching deeper levels of the skin.

20.5   CONCLUSION

In many circumstances, it may be of interest to discover whether two or more regression lines are the same. Regression lines may differ in three properties, namely, in residual variance, in slope, and in elevation, which can be tested using analysis of covariance. If there are no significant differences between regression lines, an investigator may wish to combine the data from different studies and fit a single regression line to the whole of the data.

Statnote 21

NONLINEAR REGRESSION: FITTING AN EXPONENTIAL CURVE

Nonlinear regression.

Common types of curve.

Definition of an exponential curve.

The negative decay curve.

Logarithmic transformation.

Goodness of fit to an exponential curve.

21.1 INTRODUCTION

In Statnotes 15, 17, and 18, the use of correlation and regression methods to analyze a linear relationship between two variables was described. Hence, Pearson's correlation coefficient r is used to establish whether there is a significant linear correlation between two variables. Once a linear correlation has been established, a regression line can be fitted using the method of least squares to describe the relationship in more detail. Linear regression may be adequate for many purposes, but some variables in microbiology may not be connected by such a simple relationship. The discovery of the precise relation between two or more variables is a problem of curve fitting known as *nonlinear* or *curvilinear regression*, and the fitting of a straight line to data is the simplest case of this general principle. Curve fitting may be appropriate in a variety of experimental circumstances. For example, an investigator may be interested in the pattern of decline in the number of fungal

Statistical Analysis in Microbiology: Statnotes, Edited by Richard A. Armstrong and Anthony C. Hilton
Copyright © 2010 John Wiley & Sons, Inc.

colonies with depth in the soil or the pattern of penetration of an antiseptic compound into the skin. This statnote discusses the common types of curve that can arise in microbiological research and specifically describes the fitting of a curve in which the relationship between Y and X can be described by an *exponential decay function*. Subsequent statnotes will describe the fitting of a general polynomial-type curve (see Statnote 22) and curves that require more complex fitting methods such as nonlinear estimation (see Statnote 23).

21.2   COMMON TYPES OF CURVE

There are four types of curve that commonly occur in microbiological research, and these are illustrated in Figure 21.1. The first curve is the *compound interest law* or *exponential curve* and is given by the following relation:

$$Y = a\left(\exp^{bx}\right) \tag{21.1}$$

where a and b are constants to be estimated. This is the type of curve exhibited by the growth of a bacterial population in a culture when it is increasing in numbers most rapidly. The second curve is the *exponential decay curve* where Y declines to zero from an initial value a and represents the way in which quantities often decay or decline with time:

$$Y = a\left(\exp^{-bx}\right) \tag{21.2}$$

The third type of curve is the *asymptotic curve* which increases from a value $a - b$ and then steadily approaches a maximum value a known as the *asymptote*. The asymptotic curve is given by the following relation:

$$Y = a - b\left(\exp^{-cx}\right) \tag{21.3}$$

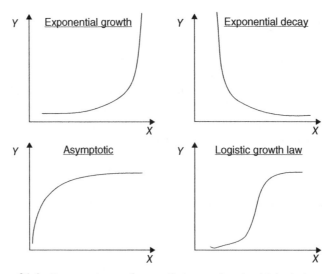

Figure 21.1. Common types of curve that occur in microbiological research.

Finally, the fourth type of curve is the *logistic growth law*, the most common sigmoid type curve and a relationship that has played a prominent part in the study of the growth of microorganisms in batch culture. Hence, the initial lag stage is followed by approximately exponential growth, but then as saturation approaches, growth slows down and reaches a maximum stationary value. The logistic growth law is given by the following equation:

$$Y = \frac{a}{1 + b \exp^{-cx}} \qquad (21.4)$$

21.3 SCENARIO

Soil has various horizons and forms a complex series of microhabitats. If soil has a uniform structure, soil organisms decline markedly within a few centimeters of the surface and continue to decrease with depth. This pattern of decline results in a typical curve exhibited by populations of bacteria in soil of uniform composition (Burges, 1967). It has been postulated that the decline in microbial numbers could be attributable to the reduction in available carbon compounds with depth, but a similar decline is also seen in peaty soils, which vary less in available carbon with depth. To determine whether soil fungi exhibit a similar curve to bacteria, the number of fungal colonies at different depths was measured in sandy soil at a site in the West Midlands, United Kingdom. The number of fungal colonies was estimated by the dilution plate method from a profile dug into the consolidated sand at the site. Samples of soil of 1 g were taken at varying depths down to 100 cm.

21.4 DATA

The data are presented in Table 21.1 and comprise measurements of the number of fungal colonies (Y) in relation to soil depth (X).

TABLE 21.1   Number of Fungal Colonies Derived by Dilution Plate Method from a Gram of Soil Collected from Different Depths in Consolidated Sand[a]

Soil Depth (cm) (X)	Colonies per Gram (Y)
0	249,000
10	110,000
20	90,000
30	60,000
40	65,000
50	50,000
60	63,000
70	45,000
80	30,000
90	37,000
100	9,000

[a] Fit to regression line of log(Y) against X: $r = -0.90$ ($P < 0.001$), $r^2 = 0.81$.

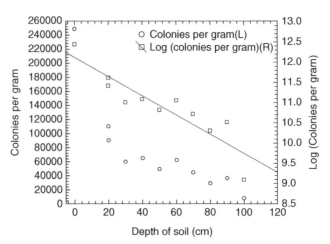

Figure 21.2. Fitting an exponential decay model to the data in Table 21.1. Circular data points represent the untransformed data and exhibit the typical features of a negative exponential curve. Square data points represent the Y values transformed to logarithms.

21.5 ANALYSIS

21.5.1 How Is the Analysis Carried Out?

To fit an exponential curve, logarithms of the Y values are taken to the base 10 and log Y is plotted against X. If the relationship is exponential (positive or negative), the graph will be linear, and a straight line can be fitted using the methods of linear regression described in Statnote 18. The method is similar to that which was used to fit a power law model in Statnote 15. Fitting a regression by transformation of the Y variable makes similar assumptions to those described for linear regression (see Statnote 18). Most statistical software will carry out this analysis as it requires only that the Y variable is transformed to logarithms and then the fitting of a conventional straight line.

21.5.2 Interpretation

The data are plotted on the original scale and on a log scale in Figure 21.2. A linear regression appears to be a good fit to the data ($r = -0.90$, $P < 0.001$, $r^2 = 0.81$), suggesting that the number of fungal colonies does decline exponentially with depth of soil, a curve similar to that shown by the soil bacteria.

21.6 CONCLUSION

Nonlinear relationships are common in microbiological research and necessitate the use of the statistical methods of nonlinear regression or curve fitting. In some circumstances, the investigator may wish to fit an exponential model to the data, that is, to test the hypothesis that a quantity Y either increases or decays exponentially with increasing X. This type of model is straightforward to fit, as taking logarithms of the Y variable makes the relationship linear, which can then be treated by the method of linear regression.

Statnote 22

NONLINEAR REGRESSION: FITTING A GENERAL POLYNOMIAL-TYPE CURVE

Do the data deviate from a straight line?
The second-order polynomial curve.
Fitting a general polynomial curve.

22.1 INTRODUCTION

An investigator may have no knowledge of the theoretical relationship connecting two variables (X, Y), but it may still be necessary to fit a curve to the data, for example, to be able to predict Y for a new value of X. If simple curvature is present, an immediate question may be whether a curve would fit the data significantly better than a straight line. Without prior knowledge of the shape of the curve, fitting a second-degree polynomial is often the best approach to this problem. Essentially, the goodness of fit of a second-order polynomial is compared with that of a straight line, ANOVA being used to test the difference between the two fits. With more complex curves, polynomials of higher order may be necessary to fit the data. Hence, polynomials of order 1, 2, 3, ... , n can be fitted successively to the data, and the addition of each extra term adds a further "bend" to the curve. Hence, third-order "cubic" curves are S shaped and fourth-order "quartic" curves have three bends and may appear to be "double peaked." An investigator may then have to decide which of the curves provides the "best" fit to the data. This statnote describes two statistical procedures: (1) to determine whether a simple curve fits better than a straight line and (2) the fitting of a more complex polynomial-type curve.

Statistical Analysis in Microbiology: Statnotes, Edited by Richard A. Armstrong and Anthony C. Hilton
Copyright © 2010 John Wiley & Sons, Inc.

22.2   SCENARIO A: DOES A CURVE FIT BETTER THAN A STRAIGHT LINE?

We return to the scenario first described in Statnote 21. Soil has various horizons and forms a complex series of microhabitats. If soil has a uniform structure, soil organisms decline markedly within a few centimeters of the surface and continue to decrease with depth. To examine the relationship between the density of soil fungi and depth, the number of fungal colonies at different depths was measured in a sandy soil at a site in the West Midlands, United Kingdom. The number of fungal colonies was estimated by the dilution plate method from a profile dug into the soil at the site. Samples of soil of 1 g were taken at varying depths down to 100 cm. In Statnote 21, it was assumed that the shape of the curve describing the decline in fungal colonies with soil depth was negative exponential. In this analysis, no prior knowledge of the possible shape of the decline in numbers with soil depth is assumed.

22.3   DATA

The data are presented in Table 22.1 and comprise measurements of the number of fungal colonies (Y) in relation to soil depth (X).

22.4   ANALYSIS

22.4.1   How Is the Analysis Carried Out?

First, a straight line is fitted to the data and an ANOVA carried out to obtain the SS of the deviation from a linear regression (see Statnote 18). A second-order (quadratic) polynomial curve is then fitted to the data, that is, an equation of the following form:

$$Y = a + bx + cX^2 \tag{22.1}$$

Hence, $Y = a + bx$ is the equation of a straight line, and the addition of the term cX^2 describes the degree of curvature. A second ANOVA is then carried out to obtain the SS of deviations from the curved regression. The difference between the linear and curvilinear SS measures the reduction in SS of the Y values achieved by fitting the curvilinear rather than the linear regression. This difference is then tested against the deviation from the curved regression using an F test. If the F ratio of the mean square reduction in SS to the

TABLE 22.1   Analysis of Variance (ANOVA) to Test Departure of Set of Data from Linear Regression[a]

Source	DF	SS	MS	F	P
Linear regression	1	26,396.509			
Second-order regression	1	32,832.989			
Reduction	1	6,436.48	6,436.48		
Deviations from second-order	8	8,685.738	1,085.717	5.93	<0.05

[a] DF = degrees of freedom, MS = mean square, F = variance ratio, P = probability.

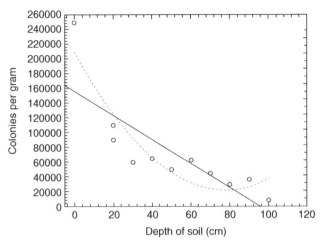

Figure 22.1. Testing whether there is a significant departure from a linear regression. Both a linear and second-order (quadratic) polynomial are fitted to the data and analysis of variance is used to determine whether a significant proportion of the remaining variance after fitting the linear relationship is accounted for by the quadratic polynomial.

mean square of the deviation from a curvilinear regression is significant, then the curved relationship is a better fit to the data than the straight line. Second-order regressions will often work well for estimation and interpolation within the range of the data even if the actual relationship between Y and X is not strictly quadratic. Extrapolation beyond the data for estimation, however, is extremely risky. If several values of Y are available at each X, then the goodness of fit of the line can be tested more rigorously (Snedecor & Cochran, 1980).

22.4.2   Interpretation

The analysis of variance of the data is shown in Table 22.1, and both regression lines are shown in Figure 22.1. The difference in the SS as a result of fitting a second-order polynomial compared with a linear regression is calculated and its mean square tested against the mean square deviation from a second-order polynomial. A value of $F = 5.93$ was obtained, which is significant at the 5% level of probability ($P < 0.05$). Hence, the second-order polynomial is a significantly better fit to the data than the linear regression.

22.5   SCENARIO B: FITTING A GENERAL POLYNOMIAL-TYPE CURVE

The human disease Creutzfeldt–Jakob disease (CJD) is caused by unusual proteinaceous infectious agents called *prions* (Will et al., 1996). Characteristic of the brain pathology of CJD is the development of vacuoles (*spongiform change*) within the cerebral cortex, resulting from the death of neurons. An investigator wished to determine if the extent of the vacuolation varied across an area of cerebral cortex from pia mater to white matter.

The specific objectives were to determine which cortical laminae were significantly affected and, therefore, which aspects of brain processing were likely to be impaired (Armstrong et al., 2002b, 2002c). To obtain the data, five traverses from the pia mater to the edge of the white matter were located at random within an area of cortex. The vacuoles were counted in $50 \times 250\,\mu m$ sample fields, the larger dimension of the field being located parallel with the surface of the pia mater. An eyepiece micrometer was used as the sample field and was moved down each traverse one step at a time from the pia mater to the white matter. Histological features of the section were used to correctly position the field. Counts from the five traverses were added together to study the vertical distribution of lesions across the cerebral cortex.

22.6 DATA

The data comprise estimates of the density of vacuoles (Y) at different distances below the pia mater of the brain (X) and are presented in Table 22.2.

TABLE 22.2   Fitting a General Polynomial-Type Curve[a]

Distance below Pia Mater (μm) (X)	Number of Vacuoles (Y)	Distance below Pia Mater (μm) (X)	Number of Vacuoles (Y)
50	3	550	22
100	31	600	20
150	29	650	25
200	28	700	21
250	25	750	19
300	22	800	17
350	19	850	25
400	18	900	14
450	16	950	16
500	22	1000	10

Source	Analysis of Variance[b]				
	DF	SS	MS	F	R^2
Total variation	19	825.8			
Reduction to linear	1	72.782	72.782		0.09
Deviations from linear	18	753.018	41.834	1.74 NS	
Reduction to quadratic	1	80.391	80.391		0.16
Deviation from quadratic	17	672.627	39.566	2.03 NS	
Reduction to cubic	1	77.004	77.004		0.16
Deviation from cubic	16	595.62	37.23	2.06 NS	
Reduction to quartic	1	327.7	327.7		0.49
Deviation from quartic	15	267.92	17.86	18.35[c]	

[a] Data are the density of vacuoles across an area of brain from pia mater to white matter in a case of Creutzfeldt–Jakob disease (CJD).

[b] DF = degrees of freedom, SS = sums of squares, MS = mean square, F = variance ratio, R^2 = multiple correlation coefficient, NS = not significant.

[c] $P < 0.001$.

22.7   ANALYSIS

22.7.1   How Is the Analysis Carried Out?

Polynomials of order 1, 2, 3, … , n were fitted successively to the data. With each fitted polynomial, the regression coefficients, standard errors (SE), values of t, and the residual mean square were obtained. From these statistics, a judgment can be made as to whether a polynomial of sufficiently high degree has been fitted to the data. Hence, at each stage, the reduction in the SS is tested for significance as each term is added. The analysis is continued by fitting successively higher order polynomials until a nonsignificant value of F is obtained. As a precaution, it is usually good practice to check the next order polynomial after a nonsignificant term has been fitted. This analysis may be included within the regression option within statistical software but in some packages can be found within the general linear modeling option.

22.7.2   Interpretation

The analysis (Table 22.2) suggests that the linear, quadratic, and cubic polynomials were not significant. However, the quartic polynomial was significant ($F = 11.67$, $P < 0.01$), suggesting a complex curved relationship between the distribution of the vacuoles and distance below the pia mater (Fig. 22.2) consistent with the vacuoles affecting specific cortical laminae. Incidentally, the fit to the fifth-order polynomial (not shown) was not significant.

There are various strategies that can be employed to decide which polynomial curve actually fits the data best, and these depend on the objective of the study. First, as each polynomial is fitted, the reduction in the SS is tested for significance. The analysis is then continued by fitting successively higher order polynomials until a nonsignificant value of F is obtained. The final polynomial giving a significant F is then chosen as the "best" fit

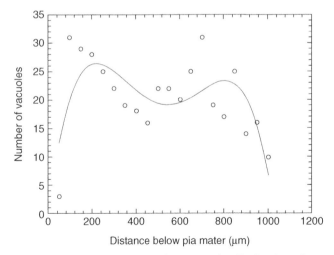

Figure 22.2. Fitting a fourth-order polynomial curve to the distribution of vacuoles in an area of cerebral cortex in a case of Creutzfeldt–Jakob disease (CJD). The distribution is bimodal with a peak of vacuole density in the upper cortex close to the surface of the pia mater and a second peak in the lower cortex.

to the data. Second, it may be obvious that a simple relationship such as a linear or quadratic polynomial would not fit the data and that a more complex curve is required. In this case, examination of the value of *multiple correlation coefficent* (R^2) (see Statnote 25) may give an indication of the correct polynomial to fit. Subsequently, F tests can be used to choose the most parsimonious model. This is the approach we adopted in the present scenario. Third, the objective may be to obtain the best possible predictions of Y from X. Hence, polynomial curves varying up to the 10th order can be fitted to the data and the curve of best fit selected on the basis of visual inspection and the highest possible regression coefficient obtained.

22.8 CONCLUSION

In some circumstances, there may be no scientific model of the relationship between X and Y that can be specified in advance, and the objective may be to provide a curve of best fit for descriptive or predictive purposes. In such an example, the fitting of successive polynomials may be the best approach. There are various strategies to decide on the polynomial of best fit depending on the objectives of the investigation.

Statnote 23

NONLINEAR REGRESSION: FITTING A LOGISTIC GROWTH CURVE

The asymptotic and logistic regression curves.
General method of fitting nonlinear regressions.
Nonlinear estimation.
Fitting a logistic growth curve.

23.1 INTRODUCTION

There are a number of curves that arise in microbiological research that cannot easily be reduced to straight lines by data transformation (see Statnote 21) or are not well fitted by a general polynomial-type curve (see Statnote 22). Examples of such curves include the *asymptotic* regression and *logistic* growth curves described in Statnote 21. Asymptotic or logistic regressions are best fitted using statistical software employing *nonlinear estimation methods*. Nonlinear estimation is a general curve-fitting procedure that can usually estimate any kind of relationship between two variables X and Y. As an example of the method, we fitted a logistic regression equation to a bacterial growth curve obtained in liquid culture.

23.2 SCENARIO

Bacteria can move freely through a liquid medium either by diffusion or active locomotion. Hence, as the cells grow and divide, they are commonly dispersed throughout the

Statistical Analysis in Microbiology: Statnotes, Edited by Richard A. Armstrong and Anthony C. Hilton
Copyright © 2010 John Wiley & Sons, Inc.

medium, which often becomes increasingly cloudy as the population grows. Hence, a few bacteria were introduced into a liquid nutrient medium and placed under optimum growth conditions. At regular intervals, a small volume of medium was removed and a count made of the cells. Plotting numbers against time enables a growth curve to be constructed for the bacterium under test. In such an experiment, cell division may not commence immediately, that is, there is a *lag phase* as the bacterium adapts to the new environment. Cells then begin to divide and grow at a rate maximal for the species under test, and this phase is known as the *exponential phase*. As they grow, however, cells use up the nutrients and produce waste products and eventually growth slows down or may even cease. The resultant curve is known as the logistic growth curve.

23.3 DATA

The data comprise estimates of the numbers of bacteria in a culture (Y) measured at different time intervals (X) and are presented in Table 23.1.

23.4 ANALYSIS: NONLINEAR ESTIMATION METHODS

23.4.1 How Is the Analysis Carried Out?

The statistical model for a general nonlinear regression can be written as

$$Y_i = f(\alpha, \beta, \gamma, X_i) + \varepsilon_i \qquad (i = 1, 2, \ldots, n) \tag{23.1}$$

where f is a regression function containing X_i and the parameters α, β, γ, while the errors ε_i have zero means and constant variance. The least-squares method can be used to estimate the parameters α, β, γ, by minimizing

$$\sum (I = 1 - n)[Y_i - f(\alpha, \beta, \gamma, X_i)]^2 \tag{23.2}$$

The problem in fitting this type of regression is the nonlinearity in one or more of the parameters α, β, γ. Note that the second-order polynomial regression $\alpha + \beta x + \gamma X^2$ is

TABLE 23.1 Number of Bacterial Colonies (Y) Derived by Dilution Plate Method Measured at Different Times (X) in Liquid Culture

Time in Hours (X)	Bacterial Colonies (divided by 10^3) (Y)
1	0.5
2	2.3
3	3.4
4	24
5	54.7
6	82.1
7	94.8
8	96.2
9	96.4

linear in the parameters α, β, γ. Hence, initial estimates a_i, b_i, c_i are made by the computer software of α, β, γ, and then Taylor's theorem is used to refine the estimates (Snedecor & Cochran, 1980). The user will normally have to specify initial estimates of the parameters defining the model. The program may also have a "stopping rule," that is, the analysis will terminate when the residual SS changes by less than a specified amount on each of a number of successive iterations.

The equation for the logistic curve is entered into the software and the least-squares method selected as the nonlinear estimation method. There are several different formulations of the logistic growth curve, one of the simplest being given by the following equation:

$$Y = \frac{b_0}{1 + b_1 \exp(-b_2 X)} \tag{23.3}$$

In Eq. (23.3), b_0 represents the upper limit of population growth, b_1 the lower limit, and b_2 the growth rate of the population. There are a number of methods available for nonlinear estimation, the least-squares method being the most widely used. Reasonable estimates of the starting values for the parameters b_0, b_1, and b_2 are also entered. Most statistical packages incorporating a general linear modeling option will be able to carry out this type of analysis. We used STATISTICA software (Statsoft Inc., Tulsa, OK) to carry out the present analysis.

23.4.2 Interpretation

The statistical significance of the parameters b_0, b_1, and b_2 are shown in Table 23.2, and the resulting fitted logistic curve in Figure 23.1, the statistical significance of the parameters indicating a very good fit to the data. This method can be used to fit any model whose mathematical equation is known and will provide estimates of the defining constants and a test of the goodness of fit of the curve to the points.

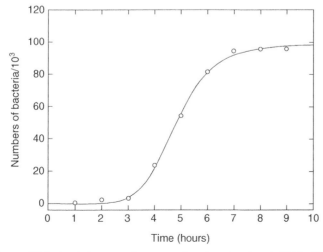

Figure 23.1. Logistic regression curve fitted to the data in Table 23.1.

TABLE 23.2   Estimation of Parameters of Logistic Curve by Nonlinear Estimation Methods[a]

Parameter	Estimate	Standard Error	t	P
b_0	99.54	1.57	63.51	<0.001
b_2	4.80	0.05	95.67	<0.001
b_1	6.76	0.42	16.02	<0.001

[a] b_0 = Upper limit of population growth, b_1 = lower limit, b_2 = growth rate of the population, $t = t$ test, P = probability.

23.5   CONCLUSION

The techniques associated with regression, whether linear or nonlinear, are some of the most useful statistical procedures that can be applied in microbiology. In some cases, there may be no scientific model of the relationship between X and Y that can be specified in advance and the objective may be to provide a *curve of best fit* for predictive purposes. In such cases, the fitting of successive polynomials is the best approach. The investigator may have a specific model in mind that relates Y to X, and the data may provide a test of this hypothesis. Some of these curves may be converted to straight lines by transformation, for example, the exponential growth and decay curves. In other cases, for example, the asymptotic or logistic curve, the regression will need to be fitted by a more complex process involving nonlinear estimation.

Statnote 24

NONPARAMETRIC ANALYSIS OF VARIANCE

Nonparametric ANOVA.
Kruskal–Wallis test for a one-way design.
Friedmann's analysis for a two-way design.

24.1 INTRODUCTION

In Statnotes 6 to 14, different types of ANOVA were described and were applied to various experimental designs. To carry out an ANOVA, several assumptions are made about the experimental data that have to be at least approximately true for the tests to be strictly valid. One of the most important of these assumptions is that the measured quantity must be a parametric variable, that is, a member of a normally distributed population. If the data do not conform to a normal distribution, then one approach is to transform the data to a different scale so that the new variable is more likely to be normally distributed (see Statnote 4). An alternative method, however, is to use a nonparametric form of ANOVA. There are a limited number of such tests available, but two useful tests will be described in this statnote, namely, the Kruskal–Wallis and Friedmann's tests.

24.2 SCENARIO

A soil microbiologist wished to determine the percentage of anaerobic bacteria present within the different horizons of a soil. Two experiments were envisaged employing two

Statistical Analysis in Microbiology: Statnotes, Edited by Richard A. Armstrong and Anthony C. Hilton

different experimental designs. In the first experiment, five random samples of soil were collected from each of five different soil horizons and designated (from the soil surface down) A1 to C. The total number of bacteria and the numbers of anaerobic bacteria were estimated within each sample by dilution plate techniques. In the second experiment, a different sampling strategy was employed and a sample of soil was collected from each of the five soil horizons on each of five separate days. In both experiments, the objective was to determine whether there was a significant difference in the percentage of anaerobic bacteria between the different soil horizons.

24.3   ANALYSIS: KRUSKALL–WALLIS TEST (FOR THREE OR MORE INDEPENDENT SAMPLES)

24.3.1   Data

The data from the first experiment comprise the percentage of anaerobic bacteria present within five randomly selected replicate samples taken from each of the five soil horizons and are presented in Table 24.1. The data, therefore, comprise a *single classification in a randomized design* (see Statnote 6). Because the data are percentages, and most of the figures are less than 10%, it is probable that the data are not normally distributed (see Statnote 4). One possibility would be to transform the data to an angular (arcsin) scale (see Statnote 4) and carry out the usual one-way ANOVA, but an alternative approach is to carry out a nonparametric analysis of the data.

24.3.2   How Is the Analysis Carried Out?

The Kruskal–Wallis test is the nonparametric equivalent of the one-way ANOVA in a randomized design (see Statnote 6) and essentially tests whether the *medians* of three or

TABLE 24.1   Kruskal–Wallis Test (Nonparametric Test for Three or More Independent Groups)[a]

		Depth of Soil Horizon		
A1	A2	B1	B2	C
2	0.6	6.3	2.0	0
1.5	0.7	6.5	2.2	0.01
2.1	0.5	6.0	1.9	0.2
1.0	0.7	6.2	2.3	0
2.0	0.6	6.1	2.2	0.01

1. Assign ranks to whole data set, ties getting mean of ranks within a run.
2. Sum the ranks for each column R, calculate the square of the totals R^2, and divide R^2 by the number of observations in each column R^2/n.
3. Add up the values of R^2/n for each column to give $K = \Sigma R^2/N$.
4. Calculate $H' = [12K/N(N + 1)] - 3(N + 1)$ where N = total number of observations ($N = 25$).
5. $H' = 22.42$ and refer to table of H' statistic. In this cases $P < 0.001$.
6. If more than three treatments or five replicates per treatment refer H' to table of χ^2 with DF one less than the number of treatments.

[a] Data are the percentage of anaerobic bacteria found in five replicate samples taken within five soil horizons.

more independent groups are significantly different (Kruskal and Wallis, 1952). To carry out the test, ranks are assigned to the whole data, regardless of group, as for the Mann–Whitney test (see Statnote 4), amending the ranks within each tied run to the mean of the ranks within the run (Dawkins, 1975). The ranks in each column are then summed and the total squared, the result then being used to calculate the Kruskal–Wallis statistic H'. Then H' is referred to a table of the H' statistic to obtain a P value. When there are more than three treatments or five replicates per treatment, however, H' can be referred to the table of the χ^2 distribution (Dawkins, 1975). Most statistical packages will carry out this type of analysis as part of the nonparametric statistics option.

24.3.3 Interpretation

In the present example, $H' = 22.416$, which is significant at the 0.1% level of probability ($P < 0.001$), suggesting highly significant differences between the medians of the five groups. More specific H_0 regarding differences between pairs of medians could then be tested using the Mann–Whitney test as a post hoc procedure (see Statnote 4). The data clearly show, however, that the proportion of organisms capable of growing under reduced oxygen supply increases with depth down to the horizon B1 but then declines below this level.

24.4 ANALYSIS: FRIEDMANN'S TEST (FOR THREE OR MORE DEPENDENT SAMPLES)

24.4.1 Data

The data from the second experiment comprise the percentage of anaerobic bacteria in a single sample taken within each of the five soil horizons but collected on five separate days and are presented in Table 24.2. The data therefore comprise a *two-way classification*

TABLE 24.2 Friedmann's Test (Nonparametric Test for Three or More Dependent Groups)[a]

| Day | Depth of Soil Horizon | | | | |
	A1	A2	B1	B2	C
1	3	0.8	7.5	2.5	0.03
2	2.5	0.7	7.2	2.4	0.01
3	2	0.6	6.3	2.0	0
4	1.5	0.5	6.0	2.2	0.01
5	1	0.5	6.2	1.9	0
Sum of ranks	17.5	10.0	25.0	17.5	5.0

1. Rank scores for each day separately; ties getting the mean of ranks within the tie.
2. Calculate the sum of ranks $\Sigma(R)$ for each column and $\Sigma(R^2)$.
2. K = number of treatments.
3. Calculate $S = \Sigma(R^2) - (\Sigma R)^2/K$ and refer to table of Friedmann's S.
4. If table of S is not available, calculate $\chi^2 = 6S/\Sigma R = 19.19$ ($P < 0.001$).
5. Refer to table of χ^2 with $K - 1$ DF.

[a]Data are the percentage of anaerobic bacteria found within five soil horizons collected on five separate days.

in a randomized blocks design, and the appropriate analysis is the nonparametric equivalent of the two-way ANOVA described in Statnote 11.

24.4.2 How Is the Test Analysis Carried Out?

As for the Kruskal–Wallis test, Friedmann's test compares the medians of three or more dependent groups (Friedmann, 1937). The scores are ranked individually for each day with tied values given the mean of the ranks as usual. The ranks are summed for each soil horizon and the square of the sum of ranks calculated. The statistic S is then obtained as shown in Table 24.2 and referred to a table of Friedmann's S or, if not available, χ^2 is calculated as shown and taken to the χ^2 table for $K - 1$ DF, where $K =$ the number of treatments, to obtain a P value.

24.4.3 Interpretation

In the present example $\chi^2 = 19.19$ and is significant at the 0.1% level of probability $(P < 0.001)$, indicating a significant effect of soil horizon on percentage of anaerobic bacteria. Examination of the data also suggests there may be some differences between samples collected on the different days, with slightly higher percentages of anaerobic bacteria recorded on the first two days. Unfortunately, Friedmann's test does not provide an explicit test of the difference between days but, as for the two-way ANOVA, takes into account variations between the different days in assessing differences between the soil horizons.

24.5 CONCLUSION

There are a limited number of nonparametric tests available for comparing three or more different groups. Two useful nonparametric tests are the Kruskal–Wallis and Friedmann's tests. The Kruskal–Wallis test is the nonparametric equivalent of the one-way ANOVA (Statnote 6) and essentially tests whether the medians of three or more independent groups are significantly different. Friedmann's test compares the medians of three or more dependent groups and in the nonparametric equivalent of the two-way ANOVA (see Statnote 11).

Statnote 25

MULTIPLE LINEAR REGRESSION

Uses of multiple regression.
Theory of multiple regression.
Multiple correlation coefficient R.
Interpretation of regression coefficients.

25.1 INTRODUCTION

In Statnotes 15 and 18, the application of correlation and regression methods to the analysis of two variables (X, Y) was described. These methods can be used to determine whether there was a linear relationship between the two variables (see Statnote 15), whether the relationship was positive or negative, to test the degree of significance of the linear relationship, and to obtain an equation relating Y to X (see Statnote 18). When the data are normally distributed, the degree of linear correlation between two variables can be tested using Pearson's correlation coefficient (r) (see Statnote 15) while if the data are not normally distributed, a nonparametric correlation coefficient can be used (see Statnote 16). This statnote extends the methods of linear correlation and regression to situations where there are two or more X variables, that is, *multiple linear regression.*

As in Statnote 18, the variables under study are referred to as Y the dependent, outcome, or response variable and X the independent, predictor, or explanatory variable. Multiple linear regression determines the linear relationship between one dependent variable (Y) and multiple independent variables $(X_1, X_2, X_3,$ etc.). Multiple regression analysis has many uses. First, it enables a linear equation involving the X variables to be constructed

Statistical Analysis in Microbiology: Statnotes, Edited by Richard A. Armstrong and Anthony C. Hilton
Copyright © 2010 John Wiley & Sons, Inc.

that predicts Y, for example, it may be necessary to predict bacterial biomass under a set of conditions specified by a series of X variables, such as pH, temperature, and amount of nutrient medium. Second, given several possible X variables that could potentially be related to Y, an investigator may wish to select a subset of the X variables that gives the best linear prediction equation. Third, an investigator may wish to determine which of a group of X variables are actually related to Y and to rank them in order of importance. For example, an investigator may wish to determine which climatic variables are most closely related to the growth of lichenized fungi in the field and in which order of importance. Multiple regression is most useful, however, in deciding whether there are any significant variables influencing Y and, therefore, should be thought of as an exploratory method, the results of which should then be tested on a new set of data and, preferably, by a more rigorous experimental approach in which the X variables are controlled.

25.2   SCENARIO

Lichens, a symbiotic association between a filamentous fungus and an alga, are often dominant in stressful environments such as the surfaces of rock and tree bark. Under these conditions, they experience extremes of temperature, moisture supply, and low availability of nutrients. As a consequence, lichens sequester a high proportion of their carbon production for stress resistance rather than for growth. Hence, as a group lichens are particularly slow growing organisms, with many species growing at less than 2 mm per year and some at less than 0.5 mm per year (Hale, 1967).

Slow growth rates and the difficulty of growing lichens for long periods under controlled laboratory conditions have made it difficult to study the influence of environmental factors on growth. In the absence of such studies, investigation of the seasonal variations in growth in the field is one method of examining the effects of environmental factors. Significant correlations between growth and climatic variables suggest hypotheses about the causal factors limiting growth that may then be tested by more controlled physiological experiments. In the present study, the radial growth rate (RGR) of thalli of the crustose lichen *Rhizocarpon geographicum* (L.)DC was measured in successive 3-month periods over 51 months in North Wales, United Kingdom. The radial growth of 20 thalli of *R. geographicum* (Armstrong & Smith, 1987) was measured at between 8 and 10 randomly chosen locations around each thallus at 3-month intervals from April 1993 to June 1997 using the method described by Armstrong (1973). Essentially, the advance of the hypothallus, using a micrometer scale, is measured in relation to fixed markers on the substratum. Radial growth in each period was averaged for each thallus and then over the 20 thalli to examine the pattern of seasonal growth. Data for 8 climatic variables were obtained from the Welsh Plant Breeding Station, Plas Gogerddan, near Aberystwyth, and included records of (1) total rainfall over each 3-month period, (2) the total number of rain days, (3) maximum (T_{max}) and minimum (T_{min}) temperature recorded on each day and averaged for each 3-month period, (4) the total number of both air and ground frosts, (5) the total number of sunshine hours, and (6) average daily wind speed.

25.3   DATA

The data comprise a single dependent (Y) variable, namely, radial growth of the lichen in each 3-month period and eight possible defining (X) variables and are presented in Table 25.1.

TABLE 25.1   Radial Growth (RGR) (*Y*) of Lichen *R. geographicum* in 17 Successive 3-Month Periods in North Wales in Relation to Eight Measured Climatic Variables (*X*)[a]

				Independent (*X*) Variables				
RGR (*Y*)	T_{max}	Air Frosts	Rain Days	Rainfall	T_{min}	Ground Frosts	Sunshine Hours	Wind Speed
0.04	7.6	8	59	207.7	6.2	16	609	8.6
0.58	20.1	0	38	306	11.6	0	237.3	8.1
0.15	10.6	3	47	317.7	5.5	33	181.9	10.7
0.18	11.6	17	51	194.5	1.9	49	171.6	11.8
0.07	8.3	8	24	97.8	6.1	33	619	8.1
0.37	19.6	0	41	287.4	11.4	1	287.4	8.7
0.33	19.4	3	68	457.7	6	25	186.9	9.3
0.17	6.8	43	44	175.8	0.26	57	276.8	8.7
0.10	13.9	2	48	295.6	6.9	19	488.5	10
0.29	17.9	0	63	328	12	0	318.4	10
0.16	10.6	17	52	318.8	4.2	41	200.6	8.7
0.35	19.8	34	49	233.4	0.87	52	217.4	8.4
0.18	14.5	2	48	197.8	7	16	521.2	8.5
0.14	18	0	42	231.1	11.3	3	495.5	7.5
0.22	10.9	13	57	463.9	5	26	223	8.6
0.20	8.9	8	72	349.8	4	17	199.2	12.4
0.34	17.9	2	31	140.3	7.2	20	220.3	6.7

[a]T_{max} = maximum temperature, T_{min} = minimum temperature.

25.4   ANALYSIS

25.4.1   Theory

When there are only two variables (*X* and *Y*), the distribution of data points in space can be represented by a two-dimensional (2D) surface, but with three variables (*Y*, X_1, and X_2) 3D geometry is required. The theory of multiple regression will be described with reference to two independent variables, but the same principles apply to any number of *X* variables. Figure 25.1 illustrates the relationship between *Y* and two *X* variables (X_1, X_2) such that any point (A) in the 3D space is defined by three coordinates (x_1, x_2, y). The relationship between *Y* and a single *X* variable is described by the *line of best fit* as determined by the method of least squares (see Statnote 18). By contrast, with two *X* variables the data are fitted by a surface or plane (the *plane of best fit*) (Fig. 25.2), which is described by the following equation:

$$Y_R = a + b_1 X_1 + b_2 X_2 \qquad (25.1)$$

where *a* is a constant, b_1 measures the change in *Y* when X_1 increases by one unit, X_2 remaining constant, while b_2 measures the change in *Y* when X_2 increases by one unit, X_1 remaining constant; b_1 is called the *partial regression coefficient* of *Y* on X_1 and b_2 the *partial regression coefficient* of *Y* on X_2. Note that lowercase letters such as *b* will be used to indicate a sample regression coefficient, which is estimated from the data, and while β indicates the population or "true" value of the regression coefficient. As in the case of a single *X* variable, the *y* values are considered to be normally distributed about

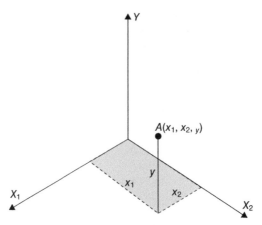

Figure 25.1. Multiple regression with two independent (*X*) variables influencing the dependent variable (*Y*). With two *X* variables, the position of any point (*A*) is described in three-dimensional space by three coordinates (x_1, x_2, *y*).

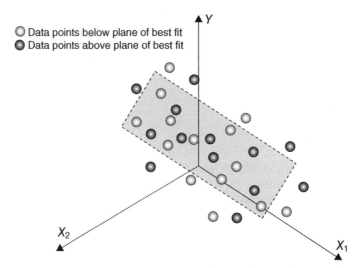

Figure 25.2. Multiple regression with two independent variables influencing *Y*. The data are fitted by the plane of best fit. The data points are scattered about the plane with some data points (*y*) above the plane and some below the plane in three-dimensional space. The degree of scatter of the points above and below the plane of best fit indicates the failure of the plane to fit the points.

the regression plane (Fig. 25.2), and the coefficients of the regression equation are chosen to minimize $\Sigma(Y - Y_L)^2$, where Y_L represents the points on the regression plane and Y the actual points. Deviations of the points from the regression plane are a consequence of random error and the existence of variables that influence *Y* but that have not been included in the study.

25.4.2 Goodness-of-Fit Test of the Points to the Regression Plane

As in the case of a single X variable, the sums of squares (SS) of the deviations of the Y values from their mean (Y^*) can be partitioned into two components, namely, the SS of the fitted points on the plane (Y_L) from their mean and the SS of the deviations of the data points (Y) from the fitted values. The goodness-of-fit test of the regression plane to the data points can then be carried out using an F test. The F test determines whether any of the X variables included in the regression are related to Y, that is, of the hypothesis that $b_1 = b_2 = 0$. Alternatively, a t test of the significance of each of the regression coefficients (b_1, b_2) can be made since ($b_1 - \beta_1$)/s_b converts b_1 so that it is a member of the t distribution with $N - k$ DF where N is the number of data points, k the number of X variables, and s_b the standard error of b_1. It should be noted that even if the regression coefficients are statistically significant, it is not uncommon that the fraction of the variation in the Y values attributable to or "explained" by the regression may be considerably less than 50%, that is, much of the variance in Y may be explained by variables that have not actually been included in the study. A multiple regression analysis in which less than 50% of the overall variance in Y is explained by the X variables will have limited value (Norman & Streiner, 1994).

25.4.3 Multiple Correlation Coefficient (*R*)

The multiple correlation coefficient (R) is defined as the simple correlation between Y and its linear regression on all of the X variables included in the study. Hence, R^2 is the fraction of the SS of deviations from the mean of the Y values attributable to the regression as a whole while $1 - R^2$ is the proportion of the SS not associated with the regression. As suggested above, a multiple regression should account for at least half the variance of the data, that is, R should be at least 0.7 ($R^2 = 0.49$). If nearly all of the variance is associated with a single X variable, a multiple regression analysis adds little to that of a simple linear regression.

25.4.4 Regression Coefficients

Multiple regression analysis is used extensively to disentangle and measure the effects of different X variables on a single Y variable. Nevertheless, there are several important limitations of this procedure especially in observational studies. In any study, there will be X variables related to Y that have not been included. These may be variables thought to be unimportant, too difficult to measure, or are unknown to the investigator. Hence, the regression coefficient of a variable, for example, b_1 is not an unbiased estimate of β_1 but of β_1 in combination with possible effects not measured. It is therefore advisable, at least initially, to include in the study all X variables that are likely to affect Y or to study a population in which variables not of direct interest can be controlled. Introducing more variables into an analysis, however, adds to the data collection effort, may contribute only noise to the prediction, and may reduce the sensitivity ("power") of the analysis. Hence, introducing new variables costs power unless each variable individually can explain important amounts of the variance. Deciding which variables to include in a study is usually a compromise between trying to achieve good predictive power while excluding irrelevant variables.

25.4.5 Interpretation

The multiple regression analysis of the lichen growth data in relation to the eight climatic variables was carried out using STATISTICA software. The value of R^2 for the present data set was 0.85, that is, the regression accounts for over half of the variance and indicates that it is worthwhile to proceed with the analysis. The ANOVA of the multiple regression is shown in Table 25.2. The value of F ($F = 5.69$) is significant at the 5% level of probability ($P < 0.05$) and, therefore, at least some of the regression coefficients are not zero. Estimates of the various regression coefficients are shown in Table 25.3. This analysis suggests that the regression coefficients for mean maximum temperature ($t = 2.43$, $P < 0.05$) and total sunshine hours ($t = -2.56$, $P < 0.05$) are the only climatic variables out of the eight to have had a significant effect on growth. Hence, lichen growth may be positively related to maximum temperature but negatively related to total sunshine hours. The likely explanation is that although warm temperatures may promote growth processes in *R. geographicum*, prolonged periods of hot dry weather actually inhibit growth presumably because of the drying out of the thalli.

25.5 CONCLUSION

Multiple linear regression determines the linear relationship between one dependent variable (Y) and multiple independent variables (X_1, X_2, X_3, etc.) and has many potential uses. An investigator should always have a clear hypothesis in mind before carrying out such a procedure and knowledge of the limitations of the analysis. In addition, multiple regression is probably best used in an exploratory context, identifying variables that might profitably be examined in more detailed studies. Where there are many variables

TABLE 25.2   Analysis of Variance (ANOVA) of Multiple Regression Data in Table 25.1[a]

Source of Variation	DF	SS	MS	F	P
Regression	8	0.2463	0.030798	5.69	< 0.05
Error	8	421.834	0.005416		

[a] DF = degrees of freedom, SS = sums of squares, MS = mean square, F = variance ratio, P = probability.

TABLE 25.3   Estimation of Regression Coefficients of Multiple Regression Data in Table 25.1[a]

Variable	b	SE of b (s_b)	t	P
Mean T_{max}	0.51	0.21	2.43	<0.05
Air frosts	0.45	0.29	1.53	NS
Raindays	−0.38	0.30	−1.26	NS
Total rainfall	0.08	0.26	0.31	NS
Mean T_{min}	−0.44	0.56	−0.78	NS
Ground frosts	−1.02	0.55	−1.86	NS
Sunshine hours	−0.57	0.22	−2.56	<0.05
Wind speed	0.006	0.22	0.03	NS

[a] b = Regression coefficient, SE = standard error, t = t test, P = probability, T_{max} = maximum temperature, T_{min} = minimum temperature, NS = not statistically significant.

potentially influencing Y, they are likely to be intercorrelated and to account for relatively small amounts of the variance. Any analysis in which R^2 is less than 50% should be suspect as probably not indicating the presence of significant variables. A further problem relates to sample size. It is often stated that the number of samples taken must be at least 5 to 10 times the number of variables included in the study (Norman & Streiner, 1994) (see Appendix 4). This advice should be taken only as a rough guide, but it does indicate that the variables included should be selected with great care as inclusion of an obviously unimportant variable may have a significant impact on the sample size required.

Statnote 26

STEPWISE MULTIPLE REGRESSION

Uses of stepwise multiple regression.
Selection of X variables for prediction.
The step-up (forward) method.
The step-down (backward) method.

26.1 INTRODUCTION

In Statnote 25, multiple linear regression was introduced as a method of studying the relationship between a dependent variable (Y) and two or more independent (X) variables. A major objective of such an analysis is often to identify the most important X variables influencing Y and to rank them in order of significance. There is usually no unique or satisfactory solution to this problem. One method would be to use the magnitude of the *standard partial regression coefficients*, usually calculated routinely in multiple regression, as a measure of the relative importance. Any ranking of the X variables, however, may be affected by correlations between the variables themselves. Multiple regression analysis assumes that the X variables are relatively independent of each other, a situation rare in practice. In addition, the contribution of a specific X variable to the total variation in Y is frequently greater when that variable is considered alone than when it is included with other variables in a multiple regression equation. For example, three different length measurements (X_1, X_2, X_3) are likely to be strongly intercorrelated and each may correlate significantly with weight (Y). If one of the X variables is entered into a regression

Statistical Analysis in Microbiology: Statnotes, Edited by Richard A. Armstrong and Anthony C. Hilton
Copyright © 2010 John Wiley & Sons, Inc.

equation, however, addition of the other two are not likely to improve the predictive power of the regression very much.

Another problem is if there are a large number of X variables included in the regression, the regression coefficients will change with each grouping of the variables. In addition, if the multiple correlation coefficient (R^2) (see Statnote 25) is small, most of the variation in Y will remain unexplained and may be attributable to random error or to variables not included in the study. Inclusion of additional variables will also change the relationships between the existing X variables and their regression coefficients. An investigator may wish to select a small subset of the X variables that give the best prediction of the Y variable. In this case, the question is how many variables should the regression equation include? One method would be to calculate the regression of Y on every subset of the X variables and choose the subset that gives the smallest mean square deviation from the regression. Most investigators, however, prefer to use a *stepwise multiple regression* procedure. There are two forms of this analysis called the *step-up* (or *forward*) method and the *step-down* (or *backward*) method. This statnote illustrates the use of stepwise multiple regression with reference to the scenario introduced in Statnote 25, namely the influence of climatic variables on the growth of the crustose lichen *Rhizocarpon geographicum* (L.)DC.

26.2 SCENARIO

We return to the scenario described in Statnote 24. The radial growth rate (RGR) of thalli of the crustose lichen *R. geographicum* measured in 17 successive 3-month periods over 51 months in North Wales. The radial growth of *R. geographicum* (Armstrong & Smith, 1987) was measured at between 8 and 10 randomly chosen locations around each thallus at 3-month intervals from April 1993 to June 1997 using the method described by Armstrong (1973). Essentially, the advance of the hypothallus, using a micrometer scale, is measured in relation to fixed markers on the substratum. Radial growth in each period was averaged for each thallus and then over the 20 thalli to examine the pattern of seasonal growth. Climatic data included records of: (1) total rainfall over each 3-month period, (2) the total number of rain days, (3) maximum (T_{max}) and minimum (T_{min}) temperature recorded on each day and averaged for each 3-month period, (4) the total number of air and ground frosts, (5) the total number of sunshine hours, and (6) average daily wind speed.

26.3 DATA

The data comprise for each 3-month period a single dependent (Y) variable, namely radial growth of the lichen and eight possible defining climatic (X) variables and are presented in Table 25.1 of Statnote 25.

26.4 ANALYSIS BY THE STEP-UP METHOD

26.4.1 How Is the Analysis Carried Out?

In the step-up method, variables are entered into the multiple regression equation one at a time. At each stage, introduction of a new variable can be tested to determine whether its effect is statistically significant, usually by an F test, or by examining whether there is a significant change in R^2. With reference to the F tests, two criteria are frequently used. First, F to enter sets an F value, which has to be exceeded before a variable will be added into the equation. Second, F to remove sets a value that after adding a new variable, a variable previously entered should be removed. If alteration in R^2 is used as the test criterion, there should be a change of at least a few percentage points before a new variable is included in the equation. The analysis continues until the next variable has an F to enter that fails to achieve significance.

In practice, the computer first calculates the regression of Y on each X separately. The X variable giving the largest reduction in the SS of deviations or the largest F to enter is selected as the first variable (X_1). Second, all the possible bivariate regressions of the remaining variables with X_1 are calculated, and the variable giving the greatest additional reduction in the SS of deviation is selected as X_2. The process is repeated with the remaining X variables until a nonsignificant reduction in the SS is achieved and the analysis is then terminated.

26.4.2 Interpretation

The analysis of the *R. geographicum* growth data by the step-up method is shown in Table 26.1. These results essentially confirm those of the original multiple regression analysis described in Statnote 25. Mean maximum temperature is the first variable selected ($R^2 = 0.57$) and, therefore, has the most important positive influence on lichen growth, accounting for 57% of the variance in growth. The second variable selected is the total number of sunshine hours during the growth period ($R^2 = 15.16$), the two variables together accounting for 72.5% of the total variance in growth, addition of the second variable increasing the explained variance by approximately 15%. The other variables were not included in the regression, having failed to meet the F to enter criterion. Hence, we would conclude that the radial growth of *R. geographicum* is most strongly related to mean maximum temperature, more weakly related to total sunshine hours, and not related to variations in rainfall or to the other variables measured.

26.4.3 Step-Down Method

In the step-down method, by contrast, the multiple regression of Y on all the X variables is calculated first. The contribution of each X to a reduction in the SS of the Y values is then computed and the variable giving the smallest reduction eliminated by a specific rule. This variable is then excluded and the process repeated until no variable qualifies for omission according to the rule employed.

TABLE 26.1 Stepwise Multiple Regression of Lichen Growth Data Presented in Statnote 25 by Step-up Method

Variable Selected	Step	R	R^2	R^2 Change	F to Enter/Remove
Mean T_{max}	1	0.76	0.57	0.57	20.17
Sunshine hours	2	0.85	0.72	0.15	7.72

[a]T_{max} = maximum temperature, R = multiple correlation coefficient, F = variance ratio.

26.5   CONCLUSION

Stepwise multiple regression techniques are useful in identifying the major variables influencing the dependent variable Y and in ranking them in order of importance. The step-up and step-down methods do not necessarily select the same variables for inclusion in the regression and these differences are magnified when the X variables are themselves intercorrelated. Investigators may sometimes use the step-up method when they wish to define the variables that influence Y more rigorously and to exclude variables that make relatively small contributions to the regression. The step-down method may retain more variables, some of which may make small contributions to the regression, but by retaining them, a better prediction may result. Investigators should also note the high probability of making a type 1 error when carrying out stepwise multiple regression, that is, claiming a significant effect when one is not present. Hence, if there were 20 X variables included in a study and none of them actually influenced Y, the probability of achieving at least one significant F to enter would be $1 - (1 - 0.05)^{20}$, that is, 0.642, greater than a 50% chance!

Statnote 27

CLASSIFICATION AND DENDROGRAMS

Purpose of classification.
Classification for "convenience" (dissection).
Theory of classification.
Methods of classification.
The unweighted pair-group method using arithmetic averages (UPGMA).
Euclidean distance.

27.1 INTRODUCTION

The analysis of bacterial genomes for epidemiological purposes often results in the production of a banding profile of DNA fragments characteristic of the genome under investigation. These may be produced using the numerous methodologies available, many of which are founded in the cutting or amplification of DNA into defined and reproducible characteristic fragments. It is frequently of interest to enquire whether the bacterial isolates are naturally classifiable into distinct groups based on their DNA profiles. A major problem with this approach is whether classification or clustering of the data is even appropriate. It is always possible to classify such data, but it does not follow that the strains they represent are "actually" classifiable into well-defined separate parts. Hence, the act of classification does not in itself answer the question: Do the strains consist of a number of different distinct groups or species or do they merge imperceptibly into one another

Statistical Analysis in Microbiology: Statnotes, Edited by Richard A. Armstrong and Anthony C. Hilton
Copyright © 2010 John Wiley & Sons, Inc.

because DNA profiles vary continuously? Nevertheless, we may still wish to classify the data for "convenience" even though strains may vary continuously, and such a classification has been called a *dissection* (Kendall & Stuart, 1966). This statnote discusses the use of classificatory methods in analyzing the DNA profiles from a sample of bacterial isolates. An approach to analyzing and representing the relationship between isolates in a nonhierarchical manner will be discussed in the final statnote.

27.2   SCENARIO

Eight unknown isolates of methicillin-resistant *Staphylococcus aureus* (MRSA) and a culture of *S. aureus* strain *NCTC8325* as a control were incubated for 18 to 24 hours at 37°C in brain–heart infusion (BHI) broth. Following incubation, the bacterial cells were harvested and 20 mg (wet weight) of cells were resuspended in 1 ml NET-100 [0.1 M Na_2EDTA (pH 8.0), 0.1 M NaCl, 0.01 M Tris-HCl (pH 8.0)] and mixed with an equal volume of molten low-melting-point chromosomal-grade agarose [0.9% (w/v) in NET-100; BioRad, UK]. The prepared blocks were incubated for 24 hours at 37°C in 3 ml lysis solution (6 mM Tris pH 7.6, 100 mM EDTA pH 8, 100 mM NaCl, 0.5% lauroyl sarcosine, and 1 mg/ml lysozyme) with 20 units of lysostaphin (Sigma, UK). The initial lysis solution was removed and the blocks were incubated for 48 hours at 50°C in 3 ml ESP [0.5 M EDTA pH 9, 1.5 mg/ml proteinase K (Sigma, UK), and 1% lauroyl sarcosine]. The blocks were washed at room temperature twice for 2 hours followed by two 1-hour washes using TE buffer (10 mM Tris and 1 mM EDTA, pH 8). A portion of each agarose block (1 × 1 × 9 mm) was digested with 20 units of *Sma*I (Roche, UK) in 0.1-ml buffer for 16 hours at 25°C. The digested DNA samples were subjected to pulsed-field gel electrophoresis (PFGE) (CHEF Mapper System, BioRad, UK) under the conditions outlined by Bannerman et al. (1995). Gels were stained with 1 µg/ml of ethidium bromide for 45 minutes and destained for 45 minutes in distilled water. Gels were visualized under ultraviolet (UV) illumination and photographed using the GeneGenius Bio Imaging System (Syngene, UK). All images were saved using the tagged image file format (TIFF). Figure 27.1 represents the resulting PFGE patterns of the eight *Sma*I genomic digests of MRSA; wells 3 and 8 carry a *Sma*I chromosomal digest from *S. aureus* strain *NCTC 8325* as a control and molecular weight marker (Murchan et al., 2003).

27.3   DATA

The data comprise the band distances migrated in millimeters from the origin of the gel for each of the DNA preparations and are presented in Table 27.1. Note that unlike the data for a multiple regression analysis (see Statnotes 25 and 26), there is no Y variable, the data comprise a series of X variables representing the bacterial strains.

27.4   ANALYSIS

27.4.1   Theory

If there are s bands defined by distance from the origin of the gel, each bacterial strain could be represented by a point in an s-dimensional coordinate frame, any one point having

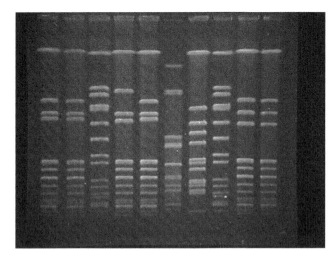

Figure 27.1. Resulting pulsed-field gel electrophoresis (PFGE) patterns of the eight SmaI genomic digests of MRSA; wells 3 and 8 carry an SmaI chromosomal digest from S. aureus strain NCTC 8325 as a control and molecular weight marker.

TABLE 27.1 Band Distances, Migrated in Millimeters, from the Origin of Gel for Each of the DNA Preparations Extracted from Several Strains of MRSA and a Culture of S. aureus Strain NCTC8325 as a Control

Band	Bacterial Strains									
	A	B	C	D	E	F	G	H	I	J
1	17	17	18	17	17	26	18	18	18	18
2	46	46	38	39	46	40	50	38	45	45
3	52	52	42	53	52	42	58	42	52	52
4	56	56	48	56	56	66	64	48	58	58
5	80	80	58	79	80	70	69	58	79	79
6	85	85	66	84	85	71	79	66	84	84
7	89	89	76	89	89	81	82	76	89	89
8	94	94	80	94	94	88	85	80	94	94
9	98	98	91	98	98	89	91	91	98	98
10	102	102	95	102	102	93	92	95	102	102
11	103	103	98	103	103	96	93	98	103	103
12	104	104	104	104	104	104	97	104	105	105
13	105	105	105	105	105		99	105		
14							109			

coordinates represented by the distances traveled by the defining bands. If in this *s*-dimensional frame, the points formed a hypersphere then the strains sampled would constitute a single homogeneous group. Alternatively, the strains may be clustered into two or more separate hyperspheres. Moreover, if the strains formed a single nonisodiametric cluster such as an ellipse, then the strains may be neither homogeneous nor divisible into separate groups but would essentially be continuously distributed without distinct "gaps."

There is a bewildering array of methods available for classifying data (Pielou, 1969). Hence, classification methods may be hierarchical or reticulate, divisive or agglomerative, and monothetic or polythetic. Moreover, several methods are available for measuring the similarity or "distance" between strains. The most commonly employed hierarchic clustering methods include nearest-neighbor, furthest-neighbor, and the unweighted pair–group method using arithmetic averages (UPGMA) (Clifford & Sokal, 1975), the most frequently used of which is UPGMA. When employing UPGMA to construct a dendrogram (tree diagram), an assumption is made during the calculation that each molecular strain type diverges equally from the others, and it is this approach that will be adopted in this statnote.

27.4.2 How Is the Analysis Carried Out?

The first problem to be considered in any classification analysis is the nature of the variables and whether the measurements have been made on the same or different scales. For example, each strain may have been defined by band distance and intensity of the band, and, if both types of data were included, the analysis would be biased by the variable with the greatest mean and range. Hence, such data are often "standardized" by converting them so that they are members of the standard normal distribution, that is, a distribution with a mean of zero and a standard deviation of one unit (see Statnote 2). Second, a distance measure d needs to be selected that reflects the similarity of one strain to another. There are various methods of computing distance, but the most straightforward is to compute d as if the s variables are dimensions making up an s-dimensional space, namely to use *Euclidean distance* as a measure of similarity. Third, a linkage rule needs to be selected, and, as discussed in Section 27.4.1, the method based on UPGMA has been the most commonly used to analyze bacterial strains. Nevertheless, note that the nearest-neighbor method is often the default option available in various statistical packages and also offers a satisfactory method of classification.

Most of the major statistical packages such as SPSS and STATISTICA offer multivariate classificatory methods, and, although the "mechanics" of carrying out the analyses may differ in detail, they use a similar approach. We will illustrate the analysis of our data using STATISTICA software.

27.4.3 Interpretation

The dendrogram obtained from the data in Table 27.1 using UPGMA and Euclidean distance methods is shown in Figure 27.2. The process of classification is agglomerative, individual strains being combined with those that they resemble most closely and then, successively, the groups are combined as the dendrogram is ascended. Hence, the data from wells *I/J*, *C/H*, and *A/B/E* immediately form three groups, the members of which are each identical according to band distances and cluster at a linkage distance of 0. Furthermore, *D* is the strain most closely related to *A/B/E*, and *F* is most closely related to *I/J*. Strain *G* appears to be the most unrelated to the others based on band distance, but at a linkage distance of approximately 125 all strains have been combined into a single group.

An obvious question to ask is how many groups should be retained? A useful additional plot in the interpretation of the data is the *graph of the amalgamation schedule* and is shown in Figure 27.3. As linkage distance increases, larger and larger clusters are formed but with a greater degree of within-cluster diversity. A clear discontinuity in the graph

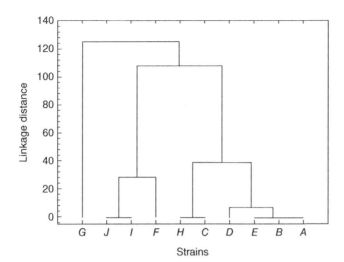

Figure 27.2. Resulting dendrogram obtained from the data in Table 27.1 using the hierarchic clustering method UPGMA and Euclidean distance as a measure of the similarity between bacterial strains.

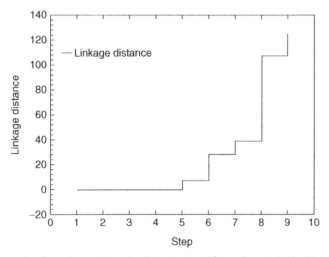

Figure 27.3. Graph of amalgamation schedule obtained from data in Table 27.1, the hierarchic clustering method UPGMA, and Euclidean distance as a measure of the similarity between bacterial strains.

suggests that many clusters are being amalgamated at the same linkage distance, and this level can be used as an approximate cut-off to determine the number of groups to retain. In Figure 27.3, for example, this discontinuity occurs at a linkage distance of approximately 40. Hence, drawing a line across the dendrogram in Figure 27.2 at this level would suggest the presence of three groups of MRSA strains, namely strain *G* alone, strains *I/J* and *F*, and strains *A/B/C/D/E* and *H*.

27.5 CONCLUSION

The use of classificatory methods is popular in the interpretation of DNA banding data from bacterial strains. The analysis results in a dendrogram that illustrates the relationships between individual strains by combining them into groups. There are two problems with this approach. First, there is no guarantee that the data are actually classifiable, and, second, there are many possible variations of the statistical analysis and their relative merits and usefulness with reference to DNA banding data have not been established. An alternative method of analysis is to make no assumptions as to the relationships between the strains. Such a nonhierarchical method of analysis involving factor analysis (FA) and principal components analysis (PCA) will be described in the final statnote.

Statnote 28

FACTOR ANALYSIS AND PRINCIPAL COMPONENTS ANALYSIS

Purpose of factor analysis (FA) and principal components analysis (PCA).

R- and Q-type analyses.

Correlation matrix.

Extraction of principal components (PC).

Stopping rules.

Factor loadings.

28.1 INTRODUCTION

In Statnotes 25 and 26, multiple linear regression, a statistical method that examines the relationship between a single dependent variable (Y) and two or more independent variables (X) was discussed. The principle objective of such an analysis was to determine which of the X variables had a significant influence on Y and to construct an equation that predicts Y from the X variables. *Principal components analysis* (PCA) and *factor analysis* (FA) are also methods of examining the relationships between different variables, but they differ from multiple regression in that no distinction is made between the dependent and independent variables; all variables are essentially treated the same. Originally, PCA and FA were regarded as distinct methods, but in recent times they have been combined into a single analysis, PCA often being the first stage of an FA (Norman & Streiner, 1994). The basic objective of a PCA/FA is to examine the relationships between the variables or

Statistical Analysis in Microbiology: Statnotes, Edited by Richard A. Armstrong and Anthony C. Hilton
Copyright © 2010 John Wiley & Sons, Inc.

the "structure" of the variables and to determine whether these relationships can be explained by a smaller number of "factors."

In Statnote 27, the application of classificatory methods (cluster analysis) to the analysis of the DNA profiles from a sample of eight unknown isolates of methicillin-resistant *Staphylococcus aureus* (MRSA) and a culture of *S. aureus* strain *NCTC8325* as a control was discussed. The most commonly employed hierarchical clustering method is the unweighted pair–group method using arithmetic averages (UPGMA) (Clifford & Sokal, 1975). The result of the analysis is a dendrogram (tree diagram) that classifies the bacterial strains into clusters. A major problem with this approach, however, is whether or not classification or clustering of the data is actually appropriate. The strains may consist of a number of different distinct groups or species, or they may merge imperceptibly into one another because their DNA profiles vary continuously. An alternative to cluster analysis is to examine the spatial relationships between the strains using a "non-hierarchical" method of analysis. This statnote describes the use of PCA/FA in the analysis of the differences between the DNA profiles of different MRSA strains introduced in Statnote 27.

28.2   SCENARIO

We return to the scenario described in Statnote 27 in which eight unknown isolates of MRSA and a culture of *S. aureus* strain *NCTC8325* as a control were studied. Eight unknown isolates of MRSA and a culture of *S. aureus* strain *NCTC8325* as a control were incubated for 18 to 24 hours at 37°C in brain–heart infusion (BHI) broth. Following incubation, the bacterial cells were harvested and 20 mg (wet weight) of cells were resuspended in 1 ml NET-100 [0.1 M Na$_2$EDTA (pH 8.0), 0.1 M NaCl, 0.01 M Tris-HCl (pH 8.0)] and mixed with an equal volume of molten low-melting-point chromosomal-grade agarose [0.9% (w/v) in NET-100; BioRad, UK]. The prepared blocks were incubated for 24 hours at 37°C in 3 ml lysis solution (6 mM Tris pH 7.6, 100 mM EDTA pH 8, 100 mM NaCl, 0.5% lauroyl sarcosine and 1 mg/ml lysozyme) with 20 units of lysostaphin (Sigma, UK). The initial lysis solution was removed and the blocks were incubated for 48 hours at 50°C in 3 ml ESP [0.5 M EDTA pH 9, 1.5 mg/ml proteinase K (Sigma, UK) and 1% lauroyl sarcosine]. The blocks were washed at room temperature twice for 2 hours followed by two 1-hour washes using TE buffer (10 mM Tris and 1 mM EDTA, pH 8). A portion of each agarose block (1 × 1 × 9 mm) was digested with 20 units of *Sma*I (Roche, UK) in 0.1-ml buffer for 16 hours at 25°C. The digested DNA samples were subjected to pulsed-field gel electrophoresis (PFGE) (CHEF Mapper system, BioRad, UK) under the conditions outlined by Bannerman et al. (1995). Gels were stained with 1 μg/ml of ethidium bromide for 45 minutes and destained for 45 minutes in distilled water. Gels were visualised under UV illumination and photographed using the GeneGenius Bio Imaging System (Syngene, UK).

28.3   DATA

The data comprise the band distances migrated in millimeters from the origin of the gel for each of the DNA preparations and are presented in Table 27.1. Note that the data comprise a series of X variables and there is no Y variable.

28.4 ANALYSIS: THEORY

The "variables" in a PCA/FA are defined by a number of criteria. If the variables are bacterial strains, called a *Q-type analysis* (Pielou, 1969), the criteria could be the distance traveled by the various DNA fragments on the gel. Hence, if there are s criteria, each strain can be represented by a point plotted in a Euclidean space defined by s dimensions, that is, each criterion can be considered as an axis orthogonal to all the other criteria and each strain as a point in this s-dimensional coordinate frame. Conceptually, in a PCA/FA, the original s-dimensional geometric frame is reduced to two (2D) or three dimensions (3D) such that the original spatial relationships between the strains are preserved as far as possible. Strictly, PCA is just one method, albeit the most common, for selecting these fewer dimensions.

A simple geometric model of the procedure is shown in Figure 28.1. Imagine that each of 21 variables were defined by only two criteria and the data plotted as a 2D scatter plot. Now suppose that we wish to simplify this 2D display by projecting the points on to a single line. In carrying out this procedure, there is an inevitable loss of information regarding the spatial relationships between the variables. The loss of spatial information would be minimized if the line was orientated through the cluster of points to preserve as much of the original spatial variability as possible. In PCA/FA, if the number of original axes is s, a smaller number of axes, or "factors," are extracted that account for significant proportions of the original variance in the data. If the extraction procedure is PCA, then the extracted factors are called principal components (PC). The first principal component (PC1) accounts for the maximum variance while the remaining PCs account for decreasing proportions of the remaining variance. The PC is analogous to the single line drawn through the points in Figure 28.1. Individual variables such as strains could then be plotted in relation to these extracted PCs. Such plots summarize the similarities and differences that exist between the variables, variables close together in the plot being more similar in the measured criteria than those furthest apart. Such an analysis can indicate whether

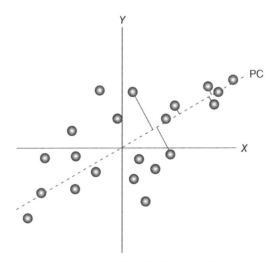

Figure 28.1. Hypothetical example in which variables are defined by only two criteria (*X*, *Y*), resulting in a two-dimensional (2D) scatter plot. In theory we could reduce the 2D display of the patients to 1D by projecting the points so that they lie on a single line (PC).

variation between bacterial strains was continuously distributed or clustered into subgroups. In other uses of PCA/FA, the variables may not be bacterial strains but the criteria used to define them (sometimes called a *R*-type analysis), and the intention may be to determine whether the variables (distances of the DNA bands) can be grouped into a smaller number of underlying factors that can best explain the patterns of variation in the data.

28.5   ANALYSIS: HOW IS THE ANALYSIS CARRIED OUT?

Most of the major statistical packages such as SPSS and STATISTICA will offer FA and PCA often as part of a *multivariate analysis* option. We will illustrate the analysis of our data using STATISTICA software.

28.5.1   Correlation Matrix

The original "raw" data set comprises the variables (the columns or strains) defined by a number of criteria (the rows or band distances). The first stage in a PCA/FA is the calculation of the correlation of each variable with all of the other variables. If the measured variables are clearly related to a smaller number of underlying factors, then this may be apparent by inspection of the correlation coefficient matrix. This is because all the variables that measure one factor would correlate strongly with each other and not with the variables associated with another factor. In practice, however, it is very difficult to extract the actual factors by visual examination of a correlation matrix since all variables are likely to show some degree of correlation with the other variables.

28.5.2   Statistical Tests on the Correlation Coefficient Matrix

The next stage of the analysis involves checking whether it is worth carrying out a PCA/FA with the existing data, that is, are there sufficiently strong correlations between the variables to analyze them factorally? The simplest of these tests involves examination of the strength of the correlations within the matrix. If there are few correlations greater than about $r = 0.30$, then it is probably not worth proceeding any further (Tabachnick & Fidell, 1989), that is, the correlations between variables are too weak for them to be combined into fewer factors. A second test involves examination of the *partial correlations* between the variables. If there are only three variables in the study (X_1, X_2, X_3), the partial correlation coefficient between any two of them, say X_1 and X_2, is the correlation between them in a cross section of individuals all having the same value of X_3. In other words, the effect of X_3 is removed from a test of the correlation between X_1 and X_2. In the case of a PCA/FA, if the variables correlate with each other because they are related to a smaller number of underlying factors, the partial correlations should be small (Guttman, 1954). Another statistical test often employed is *Bartlett's test of sphericity*, which uses χ^2 as a test statistic. If the value of χ^2 is not significant $(P > 0.05)$, then no correlations are present and the analysis should not proceed. A final test is that of *sampling adequacy* and is a measure of the degree of correlation within the data set as a whole and of the individual variables. If the sampling adequacy for the whole data set is <0.50, it is better not to proceed with the analysis. Similarly, if the sampling adequacy for an individual variable is <0.50, that variable is unlikely to show much correlation with the other variables in the study, and it may be better to proceed without it.

28.5.3 Extraction of Principal Components

The PCs are extracted so that, first, they are uncorrelated with each other and, second, that each successive PC accounts for a decreasing proportion of the remaining variance. Hence, PCA/FA tries to explain the variance of a group of variables in terms of a smaller number of uncorrelated PCs.

28.5.4 Stopping Rules

The analysis will extract as many PCs as there are variables included in the analysis. An important question is: How many PCs should actually be extracted from the data and retained for examination? There is no objective statistical method available that can determine the number of PCs. Instead, most statistical programs employ one or more rules, termed *stopping rules*, to determine how many PCs should be retained. In a PCA/FA, each PC is associated with an eigenvalue that is the percent of the total variance in the data explained by that PC. First, the *Kaiser criterion* selects all eigenvalues greater than 1. This criterion has two disadvantages: (1) It is essentially arbitrary and (2) it tends to select either too many or too few PCs depending on the number of variables in the study. Second, in *Cattell's scree test*, the eigenvalues for each factor are plotted in descending order. The change in the eigenvalue is rapid at first and then levels off, and the last factor is chosen before the flat portion of the curve. Third, the 75% stopping rule selects all PCs whose variance when added up sums to 75% and is also likely to extract too many PCs. Whichever method is used, it is commonly observed in PCA/FA that the first PC will contain a disproportionate amount of the variance and the others relatively small amounts. Hence, in many studies, it is often not worthwhile to examine more than the first three PCs.

28.5.5 Factor Loadings

The next stage involves determining which of the measured variables are associated with each PC, and this is strictly where the factor analysis part of the method begins. PCA/FA calculates the correlations between each variable and the extracted PC, and these are known as the *factor loadings*; a high value indicates a strong relationship between the variable and the PC. Ideally, each of the variables should load onto only one PC, that is, the loadings should be close to unity or zero. In practice, variables are often complex and load onto more than one PC, and this can make their interpretation more difficult. Another problem is that some variables may have positive loadings and others negative loadings on a PC. In some types of PCA/FA, rotation of the PCs is often advocated to produce a solution that is easier to interpret. Rotation can be carried out in such a way that the PCs remain uncorrelated (an *orthogonal* solution), or this condition can be relaxed and some degree of correlation between the PC can be accepted (an *oblique* solution). Rotation of the PCs will often produce a more equitable distribution of the variance between the extracted PCs and also result in individual loadings that are closer to unity or zero.

28.5.6 What Do the Extracted Factors Mean?

An important step in a PCA/FA is to attempt to interpret what the extracted PCs actually mean with reference to the problem or hypothesis posed. The first stage of this analysis involves determining which variables load significantly onto each PC. A simple procedure would be to accept as significant any variable whose loading was larger than a certain value, for example, >0.30 or >0.50; but this is an arbitrary procedure and does not take into account sample size. A more rigorous method is to test the loadings statistically using *Stevens method* (Norman & Streiner, 1994) and is given for a sample size of N by the following equation:

$$\text{Critical value} = \frac{5.152}{\sqrt{N-2}} \tag{28.1}$$

Hence, any variable whose factor loading exceeds this critical value may be regarded as being significantly correlated with a PC.

28.5.7 Interpretation

In the present application of PCA/FA, bacterial strains are the variables (Q-type analysis), and the result is a scatter plot of the strains in relation to the extracted PC. The objective is twofold: (1) to describe the pattern of variation between bacterial strains and (2) to identify those features of the DNA profiles that best correlate with the distribution of the strains.

The matrix of correlations between the MRSA strains is shown in Table 28.1. The majority of the correlations exceed 0.30, suggesting that the data are suitable for PCA/FA. Two PCs were extracted from the data, accounting for approximately 93% of the total variance in the data (Table 28.2). Hence, reducing the original 10D frame to 2D has resulted in the loss of approximately 7% of the original spatial information. A plot of the bacterial strains in relation to PC1 and PC2 is shown in Figure 28.2. The data suggest that the data from wells *I/J*, *C/H*, and *A/B/E* immediately form three groups, which are each identical according to band distances. Furthermore, *D* is the strain most closely related to *E/B/A* and *F* is most closely related to *J/I*. Strain *G* appears to be the most unrelated to the others. In addition, the correlations between the band distances for each strain and the factor loadings of the strains on PC1 and PC2 are shown in Table 28.3. Bands 4 and 14 were positively correlated with PC1 and bands 11 and 12 negatively correlated with PC1. Hence, these are the DNA band distances that are the most important in determining the clustering of the strains. Moreover, band 13 was negatively correlated with PC2. The

TABLE 28.1 Simple Correlation Matrix (Pearson's Correlation Coefficient *r*) between the Eight Bacterial Strains[a]

	A	*B*	*C*	*D*	*E*	*F*	*G*	*H*	*I*	*J*
A	1									
B	1	1								
C	0.98	0.98	1							
D	1	1	0.98	1						
E	1	1	0.98	1	1					
F	0.68	0.68	0.62	0.68	0.68	1				
G	0.48	0.48	0.52	0.49	0.48	0.20	1			
H	0.98	0.98	1	0.98	0.98	0.62	0.52	1		
I	0.70	0.70	0.63	0.70	0.70	0.99	0.24	0.63	1	
J	0.70	0.70	0.63	0.70	0.70	0.99	0.24	0.63	1	1

[a]The majority of the correlations exceed 0.30, suggesting that the data are suitable for PCA/FA.

TABLE 28.2   Unrotated Factor Loading Matrix: Correlation between Each of the Bacterial Strains and the Extracted Principal Components (PC) and Percentage of Total Variance Explained by Each PC

	Extracted PCs	
Strain	PC1	PC2
A	−0.977435	−0.161167
B	−0.977435	−0.161167
C	−0.952375	−0.255144
D	−0.977486	−0.166691
E	−0.977435	−0.161167
F	−0.804656	0.572954
G	−0.498039	−0.489264
H	−0.952375	−0.255144
I	−0.823068	0.549259
J	−0.823068	0.549259
% Variance	78.861	14.069

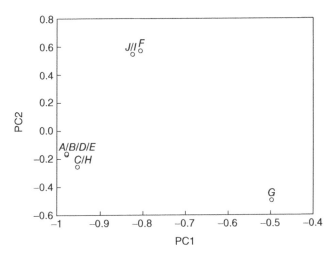

Figure 28.2. Resulting pulsed-field gel electrophoresis (PFGE) patterns of the eight SmaI genomic digests of methicillin-resistant S. aureus (MRSA); wells 3 and 8 carry an SmaI chromosomal digest from S. aureus strain NCTC 8325 as a control and molecular weight marker plotted in relation to PC1 and PC2.

interpretation of the PCA/FA data in this example is in close agreement with that of the dendrogram analysis described in Statnote 27. However, the PCA has a number of advantages over that of classification: (1) no assumptions are made that the data are actually classifiable; (2) the relationship between strains and clusters of strains is spatially displayed, which facilitates discussion of the implications of the analysis; and (3) the analysis identifies the criteria, in this case the band distances, that best differentiate between the strains.

TABLE 28.3 Correlations between Band Distances and Factor Loadings of Bacterial Strains on PC1 and PC2

Band	Extracted PCs	
	PC1	PC2
1	0.26	0.54
2	0.51	−0.08
3	0.39	−0.18
4	0.66^{a}	0.40
5	−0.13	0.32
6	−0.04	0.13
7	−0.16	0.28
8	−0.21	0.38
9	−0.42	0.14
10	−0.57	0.22
11	-0.69^{a}	0.26
12	-0.80^{a}	0.55
13	−0.31	-0.96^{b}
14	0.88^{b}	−0.44

$^{a}P < 0.01.$
$^{b}P < 0.001.$

28.6 CONCLUSION

PCA/FA are methods of analyzing complex data sets in which there are no clearly defined X or Y variables. They have multiple uses, including the study of the pattern of variation between individual entities such as bacterial strains and the detailed study of descriptive variables. In most applications, variables are related to a smaller number of *factors* or PCs that account for the maximum variance in the data and, hence, may explain important trends among the variables. No assumptions are made before the analysis that the variables can actually be classified, and this may be a considerable advantage in the analysis of more complex data sets in which DNA band data among strains may be more continuously distributed.

REFERENCES

Abacus Concepts (1993). *SuperANOVA*. Abacus Concepts Inc., Berkeley CA.

Armstrong, R. A. (1973). Seasonal growth and growth rate colony size relationships in six species of saxicolous lichens. *New Phytol* **72**: 1023–1030.

Armstrong, R. A. & Hilton, A. (2004). The use of analysis of variance (ANOVA) in applied microbiology. *Microbiologist* **5**: 18–21.

Armstrong, R. A. & Smith, S. N. (1987). Development and growth of the lichen *Rhizocarpon geographicum. Symbiosis* **3**: 287–300.

Armstrong, R. A., Slade, S. V., & Eperjesi, F. (2000). An introduction to analysis of variance (ANOVA) with special reference to data from clinical experiments in optometry. *Ophthal Physiol Opt* **20**: 235–241.

Armstrong, R. A., Eperjesi, F., & Gilmartin, B. (2002a). The application of analysis of variance (ANOVA) to different experimental designs in optometry. *Ophthal Physiol Opt* **22**: 1–9.

Armstrong, R. A., Cairns, N. J., Ironside, J. W., & Lantos, P. L. (2002b). Quantification of vacuolation ("spongiform change"), surviving neurons and prion protein deposition in eleven cases of variant Creutzfeldt–Jakob disease. *Neuropathol Appl Neurobiol* **28**: 129–135.

Armstrong, R. A., Cairns, N. J., Ironside, J. W., & Lantos, P. L. (2002c). Laminar distribution of the pathological changes in the cerebral cortex in variant Creutzfeldt–Jakob disease (vCJD). *Folia Neuropathol* **40**: 165–171.

Bannerman, T. L., Hancock, G. A., Tenover, F. C., & Miller, J. M. (1995). Pulsed-field gel electrophoresis as a replacement for bacteriophage typing of *Staphylococcus aureus. J Clin Microbiol* **33**: 551–555.

Bartlett, M. S. (1937). Properties of sufficiency and statistical tests. *J Roy Stat Soc* **160**: 268–282.

Bland, J. M. & Altman, D. G. (1986). Statistical method for assessing agreement between two methods of clinical measurement. *Lancet* **I**: 307–310.

Bland, J. M. & Altman, D. G. (1996). Measurement error and correlation coefficients. *BMJ* **313**: 41–42.

Brown, M. B. & Forsythe, A. B. (1974). Robust tests for the equality of variances. *J Am Stats Assoc* **69**: 264–267.

Burges, A. (1967). *Micro-organisms in the Soil*. Hutchinson University Library. Hutchinson, London.

Clifford, H. T. & Sokal, R. R. (1975). *An Introduction to Numerical Classification*. Academic, New York.

Cochran, W. G. & Cox, G. M. (1957). *Experimental Designs*, 2nd ed. Wiley, New York.

Dawkins, H. C. (1975). *Statforms: "Pro-formas" for the Guidance of Statistical Calculations*. Edward Arnold, London.

Dunnett, C. W. (1980a). Pairwise multiple comparisons in the homogeneous variance, unequal sample size case. *J Am Stat Assoc* **75**: 789–795.

Statistical Analysis in Microbiology: Statnotes, Edited by Richard A. Armstrong and Anthony C. Hilton

Dunnett, C. W. (1980b). Pairwise multiple comparisons in the unequal variance case. *J Am Stat Assoc* **75**: 796–800.

Fisher, R. A. (1922). On the interpretation of χ^2 from contingency tables, and the calculation of P. *J Roy Stat Soc* **85**: 87–94.

Fisher, R. A. (1925). *Statistical Methods for Research Workers*. Oliver & Boyd, Edinburgh.

Fisher, R. A. (1935). *The Design of Experiments*. Oliver & Boyd, Edinburgh.

Fisher, R. A. & Yates, F. (1963). *Statistical Tables for Biological, Agricultural and Medical Research*, 6th ed. Longman, Edinburgh.

Freese, F. (1984). *Statistics for Land Managers*. Paoeny Press, Jedburgh, Scotland.

Friedmann, M. (1937). The use of ranks to avoid the assumptions of normality implicit in the analysis of variance. *J Am Stat Assoc* **32**: 675–701.

Games, P. A. & Howell, J. F. (1976). Pairwise multiple comparison procedures with unequal n's and/or variances: A Monte Carlo Study. *J Educat Stat* **1**: 113–125.

Guttman, L. (1954). Some necessary conditions for factor analysis. *Psychometrica* **19**: 149.

Hale, M. E. (1967). *The Biology of Lichens*. Contemporary Biology Series, Edward Arnold, London.

Hilton, A. C. & Austin, E. (2000). The kitchen dishcloth as a source of and vehicle for foodborne pathogens in a domestic setting. *Int J Env Health Res* **10**: 257–261.

Katz, D. L. (1997). *Epidemiology, Biostatistics and Preventive Medicine Review*. W.B. Saunders, Philadelphia.

Kendall, M. (1938). A new measure of rank correlation. *Biometrika* **30**: 81–89.

Kendall, M. G. & Stuart, A. (1966). *The Advanced Theory of Statistics*, Vol. **3**. Griffin, London.

Keselman, H. J. & Rogan, J. C. (1978). A comparison of the modified Tukey and Sheffé methods of multiple comparisons for pairwise contrasts. *J Am Stat Assoc* **73**: 47–51.

Kruskal, W. & Wallis, W. A. (1952). Use of ranks in one-criterion variance analysis. *J Am Stat Assoc* **47**: 583–621.

Lehr, R. (1992). Sixteen S-squared over D-squared: A relation for crude sample size estimates. *Stat Med* **11**: 1099–1102.

Levene, H. (1960). In *Contributions to Probability and Statistics*. Stanford University Press, Stanford, CA.

Mann, H. B. & Whitney, D. R. (1947). On a test of whether one of two random variables is stochastically larger than the other. *Ann Math Stat* **18**: 50–60.

Murchan, S., Kaufmann, M. E., Deplano, de Ryck, A. R., Struelens, M., Zinn, C. E., Fussing, V., Almenlinna, S., Vuopio-Varkila, J., El Solh, N., Cuny, C., Witte, W., Tassios, P. T., Legakis, N., van Leeuwen, W., van Belkum, A., Vindel, A., Laconcha, I., Garaizar, J., Haeggman, S., Olsson-Liljequist, B., Ransjo, U., Coombes, G., & Cookson, B. (2003). Harmonization of pulsed-field gel electrophoresis protocols for epidemiological typing of strains of methicillin-resistant *Staphylococcus aureus*: A single approach developed by consensus in 10 European laboratories and its application for tracing the spread of related strains, *J Clin Microbiol* **41**: 1574–1585.

Norman, G. R. & Streiner, D. L. (1994). *Biostatistics: The bare essentials*. Mosby, St. Louis.

Pearson, E. S. & Hartley, H. O. (1954). *Biometrika Tables for Statisticians*, Vol **1**. Cambridge University Press, Cambridge.

Pearson, K. & Lee, A. (1902). On the laws of inheritance in man. I. Inheritance of physical characteristics. *Biometrika* **2**: 357.

Pielou, E. C. (1969). *An Introduction to Mathematical Ecology*. Wiley, New York.

Pollard, J. H. (1977). *A Handbook of Numerical and Statistical Techniques*. Cambridge University Press, Cambridge.

Ridgman, W. J. (1975). *Experimentation in Biology*. Blackie, London.

Scheffé, H. (1959). *The Analysis of Variance*. Wiley, New York.

Smith, S. N., Armstrong, R. A., & Rimmer, J. J. (1984). Influence of environmental factors on zoospores of *Saprolegnia diclina*. *Trans Br Mycol Soc* **82**: 413–421.

Snedecor, G. W. & Cochran, W. G. (1980). *Statistical Methods*, 7th ed. Iowa State University Press, Ames, IA.

Spearman, C. (1904). The proof and measurement of association between two things. *Am J Psychol* **15**: 72–101.

Spjotvoll, E. & Stoline, M. R. (1973). An extension of the *t*-method of multiple comparisons to include cases with unequal sample sizes. *J Am Stat Assoc* **69**: 975–979.

Tabachnick, B. G. & Fidell, L. S. (1989). *Using Multivariate Statistics*, 2nd ed. Harper and Row, New York.

Wilcoxon, F. (1945). Individual comparisons by ranking methods. *Biomet Bull* **1**: 80–83.

Will, R. G., Ironside, J. W., Zeidler, M., Cousans, S. N., Estebeiro, K., Alperovitch, A., Poser, S., Pocchiari, M., Hofman, A., & Smith, P. G. (1996). A new variant of Creutzfeldt–Jakob disease in the United Kingdom. *Lancet* **347**: 921–925.

Appendix 1

WHICH TEST TO USE: TABLE

The first column in the following table lists the type of data to be analyzed and the second and third columns the recommended parametric and nonparametric statistical procedures, respectively, that could be applied to the respective data. The sections of the various statnotes that describe the statistical tests are given in boldface in parentheses. An alternative method of selecting the correct test using a *taxonomic key* is presented in Appendix 2.

	Possible Statistical Procedures	
Form of the Data	Parametric	Nonparametric
A single observation x	Is x a member of a specific population (**2.5**)?	—
A sample of x values	Construct frequency distribution, calculate x^*, SD, SEM, CI (**2.4**). Is X normally distributed? (**1**)	Mode, median, 95th percentile (**4**)
Two independent samples (x_1, x_2)	Unpaired t test (**3.4**)	Mann–Whitney U test (**4.7**)
Two paired samples $(x_1 - x_2)$	Paired t test (**3.6**)	Wilcoxon signed-rank test (**4.8**)
Two sets of measurements using two methods	Test of agreement: Bland and Altman (**16**)	—

Continued

Statistical Analysis in Microbiology: Statnotes, Edited by Richard A. Armstrong and Anthony C. Hilton
Copyright © 2010 John Wiley & Sons, Inc.

Form of the Data	Possible Statistical Procedures	
	Parametric	Nonparametric
Three or more independent discrete groups $(x_1, x_2, \ldots, x_n)$	One-way ANOVA, randomized design, fixed-effects model (**6**)	Kruskal–Wallis test (**24.3**)
Three or more independent groups, random variable $(x_1, x_2, \ldots, x_n)$	One-way ANOVA, randomized design, random-effects model (**10**)	—
Three or more dependent groups in blocks	Two-way ANOVA, randomized blocks (**11**)	Friedmann's test (**24.4**)
Two or more factors, completely randomized	Factorial ANOVA, randomized design (**12**)	—
Two or more factors, (major and minor factor)	Factorial ANOVA, split-plot design (**13**)	—
Two or more factors, one of which is time	Factorial ANOVA, repeated measures design (**14**)	—
Two variances (s_1, s_2)	Compare by 2-tail F test (**8.4**)	—
Three or more variances $(s_1, s_2, \ldots, s_n)$	Bartlett's test (**8.5**) Levene's test (**8.6**) Brown–Forsythe test (**8.7**)	—
Two or more frequencies F_o (single variable)	—	χ^2 goodness of fit (**1**)
2×2 contingency table (two variables)	—	$F_e > 5$, χ^2 (**5.4**) $F_e < 5$, Fishers exact test (**5.5**)
$R \times C$ contingency table (two variables)	—	$F_e > 5$, χ^2 (**5.6**) $F_e < 5$, KS test (**1**)
Two variables (X, Y) (linear)	Pearson's r, r^2 (**15**)	Spearman's r_s (**17.5**) Kendall's τ (**17.6**) Gamma (**17.7**)
	Linear regression (r^2, ANOVA, t) (**18**)	—
Two variables (two or more samples)	Compare regression lines, analysis of covariance (**20**)	—
Two variables (X, Y) (nonlinear)	Transform to linear (**21**) Fit polynomial in X (**22**) Nonlinear estimation (**23**)	—
Y and two or more X Variables $(Y, X_1, X_2, \ldots, X_n)$	Multiple regression (**25**), R^2, stepwise procedure (**26**)	—
Several X variables (no Y)	Dendrogram analysis (**27**) PCA/FA (**28**)	

CI = confidence interval, FA = factor analysis, KS = Kolmogorov–Smirnov test, PCA = principal components analysis, r = Pearson's correlation coefficient, r^2 = coefficient of determination, R = multiple correlation coefficient, SD = standard deviation, SEM = standard error of the mean, x^* = sample mean.

Appendix 2

WHICH TEST TO USE: KEY

A taxonomic key for the identification of the correct statistical procedure. This is an alternative method of finding the correct test and relies on following a key analogous to those used in taxonomy for the identification of bacteria and fungi. Starting at (1) decide which of the alternative statements applies to the data and then follow the steps shown by the numbers in parentheses until a statnote (in bold) and an appropriate tests (in italics) is indicated.

1. The data comprise frequencies, that is, counts of specific events. (2)
   The data comprise scores, for example, abundance of microorganisms on a five-point scale. (5)
   The data are measurements, that is, continuous variables measured in units. (8)

2. Objective is a test of normality: **Statnote 1**, *Goodness of fit test χ^2 test, KS test.*
   Objective is comparison of two or more frequencies of a single variable: **Statnote 1**, *Goodness of fit test χ^2 test, KS test.*
   Objective is comparison of two or more frequencies comprising two variables. (3)

3. The data comprise a 2×2 contingency table. (4)
   The data comprise more than two rows and columns: **Statnote 5**, *$R \times C$ contingency table, χ^2 test.*

4. Expected frequencies are >5: **Statnote 5**, *2×2 contingency table, χ^2 test.*
   Expected frequencies are <5: **Statnote 5**, *Fisher's 2×2 exact test.*

5. Data comprise a single set of scores of one variable: **Statnote 4**, *Median, mode, 95th percentile.*

Statistical Analysis in Microbiology: Statnotes, Edited by Richard A. Armstrong and Anthony C. Hilton
Copyright © 2010 John Wiley & Sons, Inc.

Data comprise two groups of scores for comparison. (6)

Data comprise three or more sets of scores for comparison. (7)

Data comprise scores of two variables (X, Y) for correlation: **Statnote 17,** *Spearman's r_s, Kendall's τ, gamma.*

6. Data comprise two independent groups: **Statnote 4,** *Mann–Whitney test.*

   Data comprise two dependent (paired) groups: **Statnote 4,** *Wilcoxon signed-rank test.*

7. Data comprise three or more independent groups: **Statnote 24,** *Kruskall–Wallis test.*

   Data comprise three or more dependent groups: **Statnote 24,** *Friedmann's test.*

8. Data comprise measurements that conform to the normal distribution. (9)

   Data comprise measurements that do not conform to the normal distribution: treat as scores. (5)

9. Data comprise a single measurement of one variable: **Statnote 2,** *Is x a member of a specific distribution with known parameters?*

   Data comprise samples of measurements of one or more variables. (10)

10. Data comprise a single sample of one variable: **Statnote 2,** *Mean, SD, SEM, CV, CI.*

    Data comprise two sets of measurements for comparison. (11)

    Data comprise three sets of measurements for comparison. (13)

    Data comprise two or more variables. (15)

11. Comparison of two variances: **Statnote 8,** *Two-tail F test.*

    Comparison of two means. (12)

12. Data comprise two independent groups: **Statnote 3,** *Unpaired t test.*

    Data comprise two dependent (paired) groups: **Statnote 3,** *Paired sample t test.*

13. Comparison of three or more variances: **Statnote 8,** *Bartlett's test, Levene's test, Brown–Forsythe test.*

    Comparison of three or more means. (14)

14. Data comprise three or more independent groups (fixed variable): **Statnote 6,** *One-way analysis of variance (ANOVA), randomized design.*

    Data comprise three or more groups in a hierarchical classification (random variable): **Statnote 10,** *One-way ANOVA, random-effects model.*

    Data comprise three or more dependent groups: **Statnote 11,** *Two-way ANOVA in randomized blocks.*

15. Data comprise two or more variables with the intention of determining main effects and interactions. (16)

    Data comprise two or more variables in which the objective is correlation. (17)

    Data comprise several X variables only (no Y) and objective is either classification or determining the structure of the variables. (21)

16. Replications can be randomized to all combinations of variables without restriction: **Statnote 12,** *Factorial ANOVA randomized design.*

    Two or more variables with division into major and minor factors, one often being a subdivision of the other: **Statnote 13,** *Factorial ANOVA, split-plot design.*

    Two or more variables with one variable measuring changes with time: **Statnote 14,** *Factorial ANOVA, repeated-measures design.*

17. Data comprise two variables (X, Y). (18)

    Data comprise two or more X variables $(X_1, X_2, \ldots, X_n, Y)$. (20)

18. Objective is to test linear correlation between X and Y: **Statnote 15**, *Pearson's r.*

Objective is to fit a straight line to data: **Statnote 18**, *Linear regression.*

Objective is to fit a curve to the data. (19)

Objective is to fit a line for calibration: **Statnote 19**.

Objective is to compare two or more regression lines. **Statnote 20**.

19. To fit an exponential or negative decay function: **Statnote 21**.

To fit a general polynomial-type curve: **Statnote 22**.

To fit an asymptotic or logistic curve: **Statnote 23**.

20. Objective is to obtain an equation in the X variables to explain Y: **Statnote 25**, *multiple regression.*

Objective to determine which X's influence Y and in what order of importance: **Statnote 26**, *Stepwise multiple regression.*

21. Objective is to classify the X variables into groups: **Statnote 27**, *Dendrogram analysis.*

Objective is to understand structure of X variables by reducing the number of axes: **Statnote 28**, *PCA/FA.*

Appendix 3

GLOSSARY OF STATISTICAL TERMS AND THEIR ABBREVIATIONS

Analysis of variance (ANOVA) Analysis in which the total variance in a set of data is partitioned into components associated with different effects and then compared with each other.

Arcsin scale Percentage data are often significantly skewed when the mean is small or large and, consequently, when transformed to an arcsin scale, percentages near 0% or 100% are spread out so as to increase their variance.

Attribute data Scores that are frequencies of events, e.g., the frequencies of males and females in a hospital with a particular infection, or the proportions of resistant antibiotics to MRSA in a hospital and community environment.

Bivariate normal distribution A natural extension of the normal distribution from one to two variables and underpins studies of correlation and regression.

Bland and Altman plot A graphical method of displaying the degree of agreement between two sets of data.

Central limit theorem Means from a normal distribution (N, μ, s) of individual values are themselves normally distributed with mean μ and standard deviation $s/\sqrt{n}$, where n is the number of observations in the sample.

Chi square (χ^2) A distribution resulting from measuring the deviations of an observed frequency (F_o) from an expected or predicted frequency (F_e).

Coefficient of determination (r^2) The square of the correlation coefficient r (i.e., r^2). Measures the proportion of the variance associated with the Y variable that can be accounted for or explained by the linear regression of Y on X.

Statistical Analysis in Microbiology: Statnotes, Edited by Richard A. Armstrong and Anthony C. Hilton
Copyright © 2010 John Wiley & Sons, Inc.

Coefficient of variation (CV) The SD expressed as a percentage of the mean and provides a standardized statistic to compare two or more sets of data with different means.

Components of variance Estimates of variance (σ^2) associated with different sources of variation.

Confidence intervals (CI) The degree of error involved in estimating a population mean and often plotted as a confidence interval or error bar, and indicates the degree of confidence in the sample mean as an estimate of the population mean.

Contingency table Frequency data for two or more categories of two or more variables.

Continuous variable Nondiscrete variable measured on a continuous scale, usually with units, and in which the increments between successive measurements are essentially zero.

Correlation Describes the relationship between two variables, i.e., does one variable change in a consistent manner as the other changes?

Degrees of freedom (DF) The DF of a statistical quantity is the number of observations used to calculate the statistic minus the number of parameters that have to be calculated from the data to obtain that statistic.

Dendrogram A method of displaying the relationships between a set of variables, such as bacterial strains, in a hierarchical manner using measures of similarity and cluster analysis.

Dependent variable (Y) The Y variable in a correlation or regression study usually representing outcome or response variable.

Factor analysis Examines the relationships between the X variables or the structure of the variables and determines whether these relationships can be explained by a smaller number of factors.

Factorial design An experiment in which the effects of a number of different variables can be studied at the same time so that their interactions can be measured.

Fixed effects Where the differences between subject groups or treatments are regarded as fixed or discrete effects to be estimated.

Frequency distribution A frequency distribution is constructed by dividing the variable X into classes and plotting the number of measurements $F(x)$ that fall into each class.

Goodness of fit Measuring the degree of concordance between a series of observed frequencies with an expected or predicted distribution of results.

Homogeneity of variance Where the degree of variability between observations is similar for the different groups or treatments of an experiment.

Independent variable (X) The X variable in a correlation or regression study usually representing the predictor or explanatory variable.

Kurtosis Distributions that exhibit kurtosis are either more flat-topped than normal or have an excess of observations near the mean and fewer in the tails of the distribution than normal.

Least squares A procedure for fitting a linear regression line to data by minimising the squares of the deviations of points from the line.

Mean square (MS) The average deviation from the mean obtained by dividing the sums of squares (SS) by n or $n - 1$, also known as the variance.

Median The middle value of a variable, i.e., if all the values of x were listed in ascending or descending order, the median would be the middle value of the array.

Mode The value of a variable with the highest frequency, i.e., the maximum point of the curve.

Multiple regression Extension of the methods of linear correlation and regression methods to situations where there are two or more X variables.

Multiple regression coefficient (R) The simple correlation between Y and its linear regression on all of the X variables included in the study.

Nonparametric variable Measurements of variables that do not fulfill the requirements of the normal distribution.

Normal distribution Data that approximate closely to a bell-shaped curve (also known as a Gaussian distribution). Many measurements in the biosciences follow this distribution or do not deviate significantly from it.

Null hypothesis (H_0) By convention, hypotheses are usually stated in the negative or as null hypotheses, i.e., we prefer to believe that there is no effect of treatment until the experiment proves otherwise.

One-tail test A type of hypothesis that specifies whether a positive or a negative effect of a treatment is necessary to refute the H_0.

Paired t test Appropriate analysis when experimental subjects are, first, divided into pairs and, second, the experimental treatments are allocated to each pair independently and at random. Hence, there is a restriction in the allocation of the treatments to the experimental subjects.

Parametric variable Measurements of variables that fulfill the requirements of the normal distribution.

Partial correlation The correlation between two variables X_1 and X_2 in a cross section of individuals all having the same value of X_3.

Post hoc tests Normally carried out after an analysis of variance in which specific or all comparisons between the means are tested.

Power (P') The strength or capability of an experiment to detect a specific difference given a certain sample size and SD.

Principal component (PC) Extracted factors that account for the maximum variance in the data, each successive PC accounting for decreasing proportions of the remaining variance.

Random effects Where the objective is to measure the degree of variation of a particular measurement and to compare different sources of variation in space and time.

Randomized blocks An experiment in which experimental units are first grouped into blocks or replications and treatments are applied at random to the experimental units within each block separately.

Ranked data Where a particular attribute is scored on say, a five-point scale, e.g., the degree of reading impairment experienced by a patient as determined by his or her response to a questionnaire.

Regression coefficient (b, β) An estimate of the average change in Y associated with a unit increase in X, i.e., the slope of the line.

Repeated-measures design A special case of the split-plot type of experiment in which measurements on the experimental subjects are made sequentially over several intervals of time.

Sample mean (x*) The mean of a sample of measurements taken from a larger population. To be distinguished from population mean (μ) the mean of the population.

Skew The peak of the distribution is displaced either to the left (positive skew) or the right (negative skew), and as a result the arithmetic mean is no longer a good description of the central tendency of such a distribution.

Splt-plot design A factorial design in which one factor can be considered to be a major factor and the other a minor factor and especially when levels of the latter are subdivisions of the former.

Standard deviation (s, σ) The distance from the mean to the point of maximum slope of the normal distribution curve. Hence, the SD is a measure of the spread of the distribution measured in its original units.

Standard error of mean (SEM) The SD of the population of sample means.

Standard error of the difference between two means Measures the degree of variability of the differences between two sample means.

Standard normal distribution A normal distribution with zero mean and SD = 1 unit.

Stopping rule In a factor analysis, determines how many factors should be retained for detailed study.

Sums of squares (SS) The sum of the squares of the deviations of each observation from their mean.

t Distribution A distribution that more accurately describes the behavior of small samples from a normal distribution.

Transformation To convert the original measurements so that they are expressed on a new scale that is closer to a normal distribution than the original scale.

Two-tail test A type of hypothesis that does not specify whether an increase or a decrease in a treatment would refute the H_0.

Unpaired t test Appropriate analysis when experimental subjects are allocated at random, and without restriction, to two independent treatment groups.

Variance (s^2, σ^2) The average deviation from the mean obtained by dividing the SS by n or $n - 1$, also known as the mean square.

z Distribution A statistical table of the normal distribution, called z tables, have been calculated for the standard normal distribution and give the proportion of observations that fall a given distance from the mean.

Appendix 4

SUMMARY OF SAMPLE SIZE PROCEDURES FOR DIFFERENT STATISTICAL TESTS

Statistical Procedure	Sample Size
1. Unpaired t test	$N = (Z_\alpha + Z_\beta)^2 2\sigma^2/\delta^2$
2. Paired t test	$N = (Z_\alpha + Z_\beta)^2 \sigma^2/\delta^2$
3. One-way ANOVA	Effect size (d) $\times$ Correction factor (f) depending on distribution of means.
4. Factorial ANOVA	For main effects treat as procedure 3. For interactions: no accurate estimates possible. Use approximate solutions: 15 DF rule or $16s^2/d^2$ rule
5. Correlation coefficient (r)	$\left[\left\{(Z_\alpha + Z_\beta)\sqrt{(1-r^2)}\right\}/2\right]^2 + 2$. Approximate solution: $N = 4 + 8/r$.
6. Nonparametric statistics	Use corresponding parametric estimates + 10%.
7. Multiple regression	No accurate estimate possible. Approximate solution: minimum $N = 5 - 10$ number of X variables.
8. PCA/FA	No accurate estimates possible. Approximate solution: Minimum 5 subjects per variable if at least 100 subjects.

N = sample size, Z_α = significance level of the test, Z_β = desired probability of obtaining a significant result if the true difference is δ, σ = standard deviation of the population, δ = size of difference to be detected, d = effect size, DF = degrees of freedom, PCA = principal components analysis, FA = factor analysis.

Statistical Analysis in Microbiology: Statnotes, Edited by Richard A. Armstrong and Anthony C. Hilton
Copyright © 2010 John Wiley & Sons, Inc.

INDEX OF STATISTICAL TESTS AND PROCEDURES

The statistical tests are arranged in groups. The major groups of tests are alphabetized in bold type and individual procedures relevant to each group are then alphabetized *within* each group.

Agreement 87
"Bias" 89
Bland & Altman Plots 88, 89
Limits of Agreement 89, 90
Analysis Of Variance (ANOVA) 33, 57, 63, 67, 71, 77
One-Way ANOVA, Fixed Effects Model 33
One-Way ANOVA, Random
   Effects Model 57
Repeated Measures ANOVA 77
Split-Plot ANOVA 71
Two-Way ANOVA 63
Two-Factor ANOVA 67

Chi-Square (χ^2) Test 1, 29
Goodness of Fit Test 3
Larger Contingency (Rows×Columns)
   Table 32
Phi-Square 31
2×2 Contingency Table 31
Yates' Correction 31
Classification and Dendrograms 139
Dendrogram 142
Euclidean distance 142
Graph of amalgamation schedule 143
Unweighted pair-group method using
   arithmetic averages (UPGMA) 142
Correlation 81
Bivariate Normal Distribution 91
Coefficient of Determination (R^2) 85
Pearson's Correlation Coefficient (R) 83, 84

Data Transformation 23
Angular (Arcsin) Transformation 23
Logarithmic Transformation 23
Logarithmic Transformation in Curvilinear
   Regression 112
Square Root Transformation 23

Factor Analysis 145
Bartlett's Test of Sphericity 148
Catell's Scree Test 149
Correlation Coefficient Matrix 148, 150
Factor Loading 149
"Kaiser" Criterion 149
Oblique Solution 149
Orthogonal Solution 149
Partial Correlation Coefficient 148
Q-Type Analysis 147
R-Type Analysis 148
Test of Sampling Adequacy 148
75% Stopping Rule 149

Homogeneity of Variance 45
Bartlett's Test of Homogeneity of Variance
   For Three or More Variances 47
Brown-Forsythe Test for Three or More
   Variances 48
Levene's Test for Three or More
   Variances 48
Variance Ratio Test for Two Variances 46

Linear Regression 95
ANOVA of A Regression Line 99
Comparison of Regression Lines 105
Fitting A Regression Line By The Method
   Of Least Squares 97, 98
Prediction of Y From X 101
t Test of The Slope of The Line 100

Multiple Linear Regression 127
Goodness of Fit of Data Points to
   Regression Plane 130, 131
Multiple Correlation Coefficient (R) 131
Regression Coefficients 131
Stepwise Multiple Regression, Step-Down
   Method 136

Stepwise Multiple Regression, Step-Up Method 137

Non-Linear Regression 109, 113, 119
Exponential Curve 112
General Polynomial Curve 115
Second-Degree Polynomial 114
Non-Parametric Tests 21, 91, 123
Friedman's Test for Three or More Groups in Randomised Blocks 125
"Gamma" 94
Kendall's Rank Correlation 94
Kolmogorov-Smirnov (KS) Test 3
Kruskal-Wallis Test for Three or More Unpaired Groups 124
Mann-Whitney U Test for Two Unpaired Samples 24
Median 23
Mode 23
Spearman's Correlation Coefficient (R_s) for Correlating Two Variables 93
Wilcoxon Signed Rank Test for Two Paired Samples 25
Normal Distribution 1, 7
Arithmetic Mean (X^*, μ) 9
Coefficient of Variation (CV) 10
Confidence Interval (CI) for a Sample Mean 12
Fitting the Normal Distribution to a Sample of Data 1
Is an Individual Observation Typical of a Population (z test)? 11
Kurtosis 22
Skew 22
Standard Error (SE) of The Mean 12

Standard Deviation (SD) of Population (σ) 9
SD of Sample (s) 10
Variance 36

***Post-Hoc* Tests 39**
Bonferroni Test 42, 43
Duncan's Multiple Range Test 42, 43
Dunnett's Test 42, 43
Fisher's Protected Least Significant Difference (PLSD) 42
Games/Howell Procedure 42, 43
Planned Comparisons Between Means 40
Scheffé's Test 42, 43
Spjotvoll-Stoline Test 42, 43
Student-Newman-Keuls Test (SNK) 42, 43
Tukey-Compromise Test 42, 43
Tukey-Kramer Test (HSD) 42, 43
Power (P') 51
For Two Independent Treatments 53
In More Complex Experiments 54
Principal Components Analysis (See Factor Analysis)

Sample Size (N) 51
For Two Independent Treatments 52
In More Complex Experiments 54

The *t* Test 15
Standard Error of The Difference Between Two Means 17
The Unpaired *t* Test 16
The "Paired Sample" *t* Test 18
One-Tail Test 18
Two-Tail Test 18

Printed and bound by CPI Group (UK) Ltd, Croydon, CR0 4YY

27/10/2024

14580263-0003